Class of the Carolinas

Contents

AND SUNDRY OTHER HEROES

Dedication

This book is dedicated to King Arthur and his Knights of the Round Table. Also to the apotheosis of the queen in modern times, my own Guinevere, for whom I set out on the quest of writing this.

And in memory of my late father, Jess B. Cheatham, whose nobility of spirit — and whose gentleness in strength — helped me know those same traits of character when on my quest I encountered them among heroes, great and small.

Mike Cheatham
Asheville, North Carolina

Prologue

I don't think I started out to write this collection. But legends of heroes — in sports, in politics, in daily life — have been a valued part of my legacy. I realized that when I prepared the first of the series, "The Hurricane Remembered," for *The Toccoa (GA) Record* in 1982.

That, as well as a family tradition of storytelling, played a part, I am sure, and in this latter-named consideration my late father's was a strong influence.

In the profile on Davidson College's Joseph H. McConnell, I quote Bernard Malamud's rationale for having heroes in the first place. The author of *The Natural* wrote, "Without heroes, we're all plain people and don't know how far we can go."

Me, I have gone very far indeed. Geographically, that is. In pursuit of advancement and a livelihood, I have made some 28 household moves in 33 years. But I always looked home for my heroes.

In these pages I have written about some heroes here in my own back yard. All but a handful of the profiles are about those in Western and Piedmont North Carolina, as well as the Upstate of South Carolina.

In doing so, I have passed over some favorites from my native Georgia. People like Frank Sinkwich, Charlie Trippi, and Coach Wallace Butts. John Mize, Spurgeon Chandler, and Herb Maffett. And Dan Magill, as well. But I have come to respect and admire the sports heroes right here in the Carolinas.

Those profiled here are very real to me. They are not like Red Smith's heroes, the ones he wrote of in *To Absent Friends.*

My heroes are there for me every day. Their memories and their deeds continue to sustain me.

I was once told that sports fans are most vulnerable to the *Ubi Sunt* theme. Translated from the Latin, it means something like "Where are they now?" Then I was told of the line from the French poet, Francois Villon: "Where are the snows of yesteryear? . . ."

One of the great sports writers of our time, Furman Bisher of *The Atlanta Journal-Constitution,* used to use that device. Asking, "What ever happened to . . . ?" he would supply the latest on a Tubby Walton or a Mason Leeper. I always loved it when Furman did that.

Well, who knows, you may find here an account of one of your own old heroes. I hope so, because that would make me feel good.

Those heroes of ours, they have in fact shown us how far we can go, haven't they?

Mike Cheatham

Acknowledgements

Some special people have made me realize I had these stories in me and needed to put them down. At the top of the list are Nath and Mary Thompson of Brevard, North Carolina.

I told them about my fledgling sports commentary series on a radio station, WNCW-FM, in Spindale, North Carolina, called "Heroes: Why Not the Best?" They listened to the commentary on Earl (Junior) Wooten, liked it, and Mary wrote out the check to underwrite the series.

Forty-one years ago, at a fraternity party Nath rushed me, as did Mary Brumback from Macon, the chapter sweetheart. They've been steadfast friends all that while.

Of course, I joined that fraternity. Nath and Mary made me feel good about myself then, as they do now. Theirs is a marriage that brings out the best in their friends in so many ways.

At that same party was a lanky old boy from my hometown of Toccoa, Georgia. He was the only son of the town's beloved physician, the late John H. Terrell. Always a leader, John H. Terrell, Jr. — or "The Kid," as he still is called — was a fine all-around athlete at Toccoa High School and Darlington School for Boys. He is an even finer friend.

In 1953 while in college, I lost out on my summer job at Pepperell Mills. The textile plant just outside our new home — Rome, Georgia — was out on strike. It was then that I walked in to WLAQ in that town, declaring myself a radio announcer.

Before the summer was over, I had moved to WROM-TV, creating a show on Georgia Football with the help of Dan Magill, at the time the school's great sports information director. A kind of keeper of the flame at Athens, he knows more about what goes into the making of a hero than anyone I ever met.

In this current endeavor, Roger Crawford of Rutherfordton, North Carolina, has been in my corner and has encouraged me to be better than I thought I was capable of being.

The Shelby (NC) *Star's* Tom O'Neal, Alan Ford, and Michael Howard edited the radio commentaries, and through their alchemy weekly columns somehow appeared.

The older sister of my greatest hero — my late father, Jess B. Cheatham — was Miss Montie Cheatham. Until a couple of years ago, she had always lived at the old homeplace in Jefferson, Georgia. But a series of strokes finally brought down this 102-year-old lady. Nothing else could.

Finally, there is the person I call "The Miracle of Dallas." A talented seventeen-year-old writer, athlete, and scholar, she is my only child. When I was 43 and living in Dallas, Texas, I learned Dallas Cheatham was on her way. "Dassy" doesn't like to lose, and she won't let her dad be a loser, either.

"They can't beat you, Dad; they just can't beat you," she once said when I tried to give up.

How much I owe these heroes and heroines of mine. This is my way of thanking them. Them, and the people like them who may have brought out something heroic in your life.

Mike Cheatham

Photo Credits

With any such undertaking as the writing of this volume, much credit goes to those who furnished photographs of those profiled.

In addition to the individuals (and their family members) who provided photos, our thanks go out to the following: Sports Information Departments of North Carolina State University, The University of North Carolina, Furman University, Davidson College, Notre Dame University, Clemson University, University of North Carolina at Asheville, University of Georgia, Wake Forest University, Duke University, Western Carolina University, Ohio State University and Marshall University.

Also, we thank: Asheville School, The Kansas City Chiefs, Presbyterian College, The Southern Conference, University of North Carolina at Charlotte, and T.W. Services, Inc. Too, the photo of South Atlantic League President John Moss was provided by photographer John R. Craig of TAM Productions of Lenoir, North Carolina.

Profiles

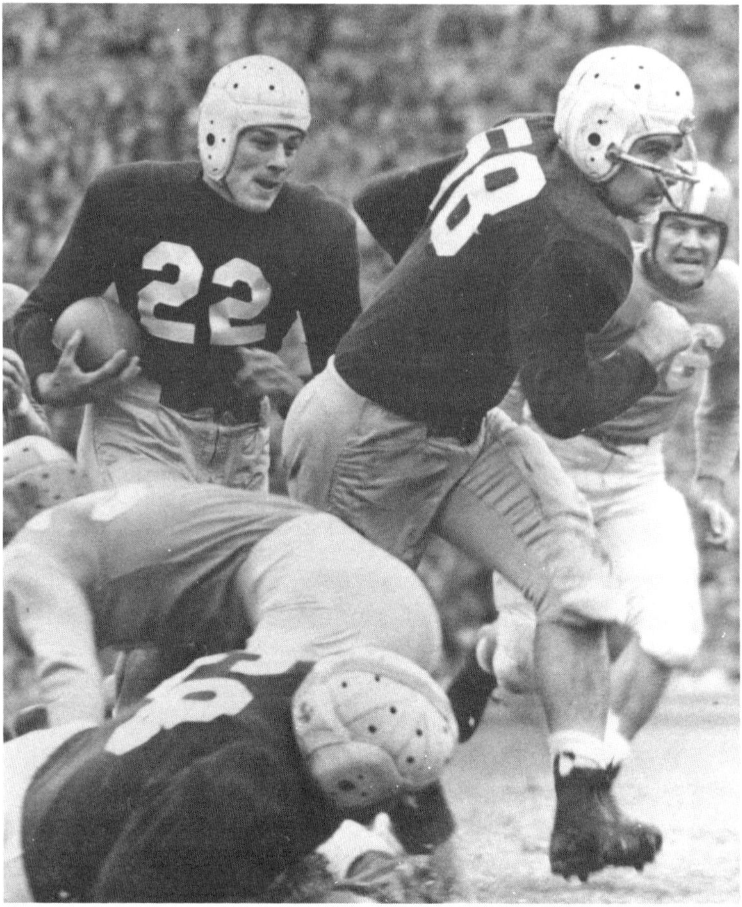

Charlie "Choo Choo" Justice was the first athlete inducted into the North Carolina Sports Hall of Fame, garnering more votes than the other three inductees combined. A two-time football All-American at Chapel Hill, he has endured as many Tarheels' all-time favorite sports hero. (Photo: Hugh Morton)

Charlie Justice

We think we know all there is to know about Charlie "Choo Choo" Justice: The football legend, the businessman, the good neighbor. The institution he has become in our state.

But we don't, not really.

Oh, the legend of the old tailback in his playing days at Chapel Hill, that's secure, a given. He was a two-time All-American, runner-up to Heisman Trophy Winner Doak Walker of SMU in his final season, 1949. He led the Tarheels to three major bowls and was MVP for the college all-stars in their upset win over the NFL Champion Philadelphia Eagles.

Charlie's handsome visage even graced the cover of America's most-read magazine, *Life*. He netted an all-time Tarheel record, 4,883 yards over a four-year span. Then a four-year career with the pros featured ample highlights, as well.

One of my colleagues at *The Star* remarked a little while back, "Doing a piece on Choo Choo, huh? I suppose maybe because he's so accessible, living over in Cherryville, we kinda take him for granted." Perhaps some folks do, but I never could.

When I resettled in these parts six years ago, in fact, I began to ask questions about Charlie Justice, The Man. Not Justice, The Legend.

What has emerged from that is a conception of the treasure the man represents, the influence for good he ever so quietly wields. His former teammates attest to that. Former Tarheel center Joe Swicegood of Asheville tells me there was never a more unselfish or cleaner player. His blocking back, Joel Wright of Hendersonville, echoes the sentiment.

I learned too of an instance where Choo Choo lent a hand to another former athlete from the area. He was an individual, down on his luck, who just needed a second chance. Ever so quietly, Charlie Justice saw to it that the man did get another chance.

Hugh Morton — a treasure of sorts on his own as owner/operator of Grandfather Mountain — gave me some other insights into Justice, the human being. Morton wrote, "He is one of my dear friends, and no one will ever say all the good things that can be said about him. If I were doing a present-day piece on him, the thing that stands out about him is the model citizen he has been since playing days.

"Charlie has lent his name to every worthwhile cause you can think of: Easter Seals, mental health, multiple sclerosis, drug rehabilitation. . . . He is just as great a citizen as he was an athlete."

The two became fast friends in 1946, Morton returning from World War II, in which he served as a combat photographer. Resuming his studies at UNC, Hugh Morton turned to a more pleasant pursuit: snapping some of the most memorable photos of the exciting Justice era. One of his classics accompanies this profile.

"The Linville Legend," Morton added this note to the letter he sent me: "I was on the original board of the North Carolina Sports Hall of Fame, and Charlie was the first person voted in, receiving more votes than Wallace Wade, Ace Parker, Jim Beatty, and Estelle Lawson Page, fast company for anybody to be in."

What's this, Justice outpointing the Duke men — Wade and Parker? Yes, he had a habit of that. During his time at Carolina, Choo Choo led the 'Heels to victory over the

Blue Devils on four straight occasions. And on November 23 of this year, Charlie is likely to be in the stands for his 45th Duke-Carolina clash.

As he thinks his own private thoughts about the contest taking place on the field, I wouldn't presume to guess what they might be. But I just hope he realizes what he has meant to us.

Charlotte Observer columnist Ron Green probably said it best about the immortal who lives so near to us here in Shelby: "(S)till the most recognized, the most revered and beloved athlete this state has ever produced . . . Justice endures as a hero of almost mystic quality."

The tribute went on, "It has its sweetness and its burdens, but neither has changed him. Nobody ever wore the mantle of an athletic hero with more style and grace than Charlie Justice."

Take this hero for granted? Not me, and I hope not you.

Fall 1991

Jim Beatty

Jim Beatty. Recognize that name?

Well, we should. He belongs to us.

The smallish native of Charlotte flashed his brilliance on an international scale after setting school records for the one-mile and two-mile run at UNC-Chapel Hill in the 1950's.

Then this personable new grad went West to train under the great Hungarian coach, Igloy. Jim Beatty had made a commitment to conditioning, excellence, and improved technique.

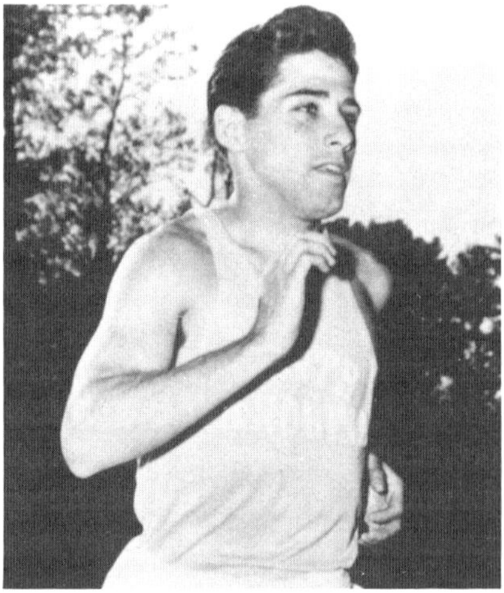

Jim Beatty was the first American to crash the indoor four-minute mile run barrier. Countless other records were owned by the single-minded competitor.

In time, he blazed into international prominence when he became the first human to break the barrier of a four-minute indoor mile. That was on February 10, 1962, and his clocking was 3:58.9. Soon came a world's record for the two-mile run at 8:29.8.

By September of 1962, Jim Beatty had broken U.S. records in five straight meets in one week — and six over a two-week period. At one stretch Beatty held 10 American and three world records at the same time.

That was all in 1962, the year when the Tarheel State's own was awarded the Sullivan Trophy as the top amateur athlete in the country.

I emphasize the word "amateur" because that's what Beatty was. He shakes his head sadly now when he sees world competition increasingly featuring subsidized athletes. The purity is gone, says the graduate of Charlotte's old Central High School.

When he left the world of big-time competition, Jim approached business and politics with the same dedication he reserved for his sport. His old coach at Chapel Hill, Dale Ranson, saluted Beatty for possessing these attributes: "Desire, competitiveness, willingness to work, and sacrifice."

No wonder, then, that Beatty has been successful both as a state legislator — he did lose a U.S. Senate race — and as the founder of his own executive search firm.

Also, just to keep his hand in, the old miler often appears on network and local television when track and field is being covered.

Jim Beatty is as trim as ever. And when you hear his voice over the telephone, it fairly crackles with energy and brims over with goodwill.

A Tarheel success story, that's Jim Beatty.

Spring 1991

Bobby Bell

Bobby Bell initially was UNC head football coach Jim Tatum's choice to break the color line in athletics in the state. Instead he went on to win the Outland Trophy at Minnesota and, after graduation, attained all-pro status with the NFL's Kansas City Chiefs.

Let's face up to something, those of us who grew up with our sports heroes in 1940's and 1950's. As heroic as our idols were, they were — by societal design — of a single race, white.

J.C. Caroline, an All-American and All-Pro from South Carolina, went to college at Illinois. George Webster, another player with those credentials, went to Michigan State.

Over in Cleveland County, Shelby's Bobby Bell signed with Minnesota. Four years later he won the coveted Outland Trophy as the nation's outstanding lineman and was third in Heisman balloting. Bobby was with the Kansas City Chiefs for 12 years, and in eight of those years he was an All-Pro linebacker.

Oh, I know, that kind of exodus doesn't happen anymore. It's pretty rare that we lose one of our bluechippers — who happens to be black — to The Big Ten or Notre Dame.

The nagging thought is still there, though. What could our home state teams have done in national competition, given the availability of talent such as Bell and his ilk?

I can't help but think about my own hometown in the early forties. My brother was the high school's captain and my dad was the coach.

J.B. confided one Saturday morning at season's end that he and his title-winning teammates were to take on a team that day from the other side of town. The contest was off-schedule and the location was off-the-record.

No, I couldn't come. Yes, it could be that Hilton, the school custodian, might know something about it. And, yes, Hilton and his younger brother, Dan, might be playing. (Months earlier, the whole THS team had begged Dad to get Hilton an athletic scholarship to Atlanta's Morris Brown College. This gentle giant of a man became a Methodist bishop.)

Close to nightfall J.B. came limping in after the epic contest. "Y'all won, didn't you?" "Ah, shaddup, kid," was the muffled reply.

But about Bobby Bell from Shelby. He lives in Kansas City now, has since retired from the Chiefs. But Coach Hank Stram paid him as fine a compliment as I ever heard.

The coach said Bell "could have played any position on the field, and your team would have won. At various times, he was a running back, tight end, quarterback, defensive tackle, defensive end, linebacker, and he snapped for us."

More stats on Bell? No, that's not the point of our story. The point is, couldn't we have used some of that versatility and talent right here in North Carolina? The late Jim Tatum wanted Bobby at Carolina but felt the pressure would have simply been too great at that time.

Bobby Bell's mother and dad still live in Shelby. "Bobby'd be proud to talk with you," says Mama last week. Seems he has been in the restaurant business for 15 years. And he wonders out loud if his hometown could use another good cafe and would I ask his friends Charlie (Choo Choo) Justice and banker Ed Hamilton what they think.

You know, I once lived in South Texas, and I loved it. Down there they learned something about what's called "resource recovery." One thing they recovered was oil. Lots of folks, in fact, call it "black gold." Yep, black gold.

In the Deep South it took us a while before we recovered one of our great resources. But Bobby Bell and I agree, I think, that we're getting there now.

Winter 1991

Fred Brown

The man works weeknights at a radio station with one of the largest shares of audience in the U.S., WWNC-AM of Asheville, North Carolina. Eleven pm to 5 am, that's his shift.

I've worked that shift in radio, and — believe me — it's lonely. But you do get some interesting (and goofy) telephone calls. Mostly from people who, well, are just as lonely as you are.

I call Robert Frederick Brown, Jr., "The Night Mayor of Asheville." Sportscaster, country and western deejay, father confessor — he's all that. Holding now at about the seven decade mark, the heart of Fred Brown is that of the lonely hunter.

A hunter, you say? What things does Brown hunt? Well, I am one of the fortunate ones, for I think I know. Oh, he hunts for peace of mind. And often, in his own quiet way, for dignity of purpose in the chambers and recesses of the hearts of those searching for "a walk in the spirit" with a higher power.

The son of a Presbyterian minister in Depression-era West Virginia, young Fred Brown grew up listening to clear channel KDKA Radio in Pittsburgh, a giant of electronics that happened to be the voice of the Pirates baseball team. At this stage of his development, too, Brown grew to admire the broadcast styles of Don Wilson and Ken Carpenter, who aired descriptions of the gridiron contests from "The Granddaddy of Them All," The Rose Bowl in Pasadena, California.

("Back then the Rose didn't have one of those infernal tie-ups with a couple of college athletic conferences — they simply invited the two best teams in America," says the announcer.)

So, a sports announcer for 46 years now, The Bearded Sage of the Blue Ridge has reached tens of thousands of fans over that span with a voice as clear and commanding as those of his boyhood idols, Wilson and Carpenter — not to mention other favorites such as Mel Allen, Ted Husing, and Red Barber.

Then why not the networks, the big time? Well, he had his chances, but an inner voice told Fred to stay here, right here in these majestic Carolina mountains. There was a daughter to raise, friends to care for, and a host of other personally compelling reasons only Brown knew.

Without going into all of that, I do know Fred Brown's deep personal faith has seen him through. His credo, recited to me at the burger joint one Sunday after church services, says much about this Christian man. It's from the book of Romans, ". . . (W)e know that all things work together for good to them who love the Lord, who are called according to his purpose."

The veteran of the airwaves, however, goes before his weeknight audience on the station for the last time on the evening of Friday, February 14, 1992. That's Saint Valentine's Day, isn't it? "All things work together for those who love the Lord. . . ."

We loved hearing Fred Brown over the years. Years when he interviewed the likes of Charlie "Choo Choo" Justice, CBS-TV's Charles Kuralt. Years when he encouraged a young Charlton Heston, then a theater group's director in Fred's town who would drop by the studio to share his dreams with the avuncular Brown.

"Oh, I'll still be handling some on-air assignments for the station from time to time," says the sage. And I tell him that maybe CBS-TV will pick him up, just like they did announcer Ernie Harwell after his employers of some 40 years, the Detroit Tigers, failed to renew the popular Harwell's contract. Fred just chuckles.

But one thing I know, this man has meant a lot to folks who tuned in late at night. Folks who tuned in for the ball scores and those "somebody-done-somebody-wrong-songs." Folks to whom, on occasion, Fred Brown supplied some raw courage. That's a noble thing to do on the lonely night air for those whose hearts are troubled.

We'll miss Fred Brown's voice on those lonely nights, won't we?

Winter 1992

Smoky Burgess

Smoky Burgess died Sunday before last as his league-leading Atlanta Braves won a big one over the second-place Los Angeles Dodgers. The new "Miracle Braves" — on top for now, as this is being written, after having finished last in 1990 — were rehabilitated and inspired by the Forest City resident, a 14-year veteran with the organization.

The typical sports page piece on his passing dwelt on the man's 18-year playing career — one in which he compiled a lifetime .295 batting average while playing for the Pirates, Phillies, Reds, and White Sox. Sadly, less was made in such stories of what Smoky felt was his greatest contribution: Developing the team's young talent, first as minor league batting instructor, then as a scout.

It so happens that my very first column for *The Star* was on Burgess several months ago. A couple of oldtimers had told me that if I wanted to do a column on Smoky, I'd best get a move on. Cancer was destroying the great catcher. These same friends of Smoky's also told me I could leave a message for him at Smith's Drug Store, that he'd make time to visit with me.

The pharmacy on East Main Street was a favorite hangout for Burgess and his many friends. It was the site for conduct of business related to the church he served so faithfully, Holly Springs Baptist. Of course, they talked a lot of baseball over there too.

Let it be known, as well, that this fellow was a Gideon

— you know, they're the people who donate Bibles to be placed in hotel rooms. I could see a young Smoky Burgess turning to one of those Bibles while worrying about a bad night at the plate after a road game, couldn't you?

"Mr. Burgess, what time should I come by?" I asked. "How about right after my chemotherapy?" (That seemed to me the very worst time.) "Gets my mind off all that when I can talk baseball with a fellow like you," said Smoky.

Smoky talked about his own career some. But he preferred talking about the other great ones. People like Roy Campanella of the Dodgers and Andy Seminick of the Phils, catchers themselves. And his great pals Gaylord Perry of Gaffney, Ewell Blackwell and Gil Coan of Brevard.

About his off-field career with Atlanta, now that's what Smoky said brought him greatest satisfaction. "I go into the locker room there now and almost every kid on the roster thanks me for helping him out on the way up — for signing him to a contract or teaching him to hit like a major leaguer."

It was a dreary, cold day in Forest City at Burgess' place on Caroleen Road. But the prospect of spring training around the corner, that warmed Smoky's heart.

The old pinch-hit champ went to the Braves camp in Florida one last time to see his boys, but he was forced to come home in a few days after the pain set in again.

But I know he was proud of the way the team's young talent had been developed painstakingly in the farm system. "This is the year," Smoky seemed to be telling me. And maybe it is. Let no one forget, though, that in a very real sense these players are "Smoky's Boys."

Wouldn't it be perfect if the club could dedicate a pennant-winning year to this man of such high character and steadfast faith? That's the way Smoky Burgess lived his life over there in Rutherford County: As a champion who went out like a champion.

Fall 1991

Dave Carey and J.L. Suttle

Dave Carey (pictured) and J.L. Suttle have been friendly rivals in Southern tennis seniors play. Going strong in the 80 and over bracket, the pair also have been over-achievers off the court.

Just ahead is what can be called "The Decade of the Eighties" for amateur tennis in these parts. And for the rest of the nation, as well.

No, I've not lost all track of time and calendar. The "Eighties" in this case means the tennis classification for competitors eighty years of age and over.

J.L. Suttle of Shelby, many times a champion in other classifications, is poised to assume the same kind of position of dominance that has been his in earlier incarnations on the court. Dave Carey, only about a year away from moving into the same bracket as Suttle, relishes the prospect of competing with the Shelby insurance executive and banker in The Eighty and Over.

What's most remarkable about this particular niche is that as late as 1965 seniors play ended with the 45 and over category. But Americans' emphasis on health and fitness — and breakthroughs in medical science — have had much to do with the growing popularity of tennis among seniors.

Take Dave Carey. A Rhodes Scholar from the University of British Columbia, he has had two hip replacements for an arthritic condition. "Each time I said to myself, 'You're going to be back on those courts in three months winning tournaments.' " Carey adds, "If you think you're an invalid, you are!" That reaction, of course, is not unexpected from one who performed for Canadian National Teams in rugby and cricket.

Winning is what the Asheville resident does, and he's the current 75 and over doubles king of North Carolina. And in the 1980's after a heart attack, the bearded netter came back to rank number one in the Southeastern states in singles in his classification.

Also overcoming the effects of a coronary, Suttle won the 75 and over singles for the state in '89 and '90. He won in the 80 and over this year, picking up the trophy as well in the Southern.

Both men express admiration for their peers on "the seniors circuit." People like Vince Connerat of Charlotte, Jim Hamilton of Asheville, and Jack Staton of Atlanta. Add to those worthies, in other classifications, Ed White of Davidson, North Carolina, and Dan Magill of Athens, Georgia.

Another shared attribute for these veterans is a heightened sense of community service. For instance, Suttle and business partner Buck Archer established the Shelby Lawn Tennis Association, and it has flourished under their care.

Dave Carey serves his town by teaching at "The College for Seniors" at UNC-Asheville, and reads for the blind on a local radio station. He has done a 26-episode series on public television, "The 100 Key Events in History."

Typical of Carey, he and his wife will go bobsledding and snowshoeing in Quebec over Christmas this year. ("Just always wanted to.") What'll Suttle be doing over the holidays? Guess he'll be readying himself for the inevitable challenge from Mr. Carey.

Over in Carey's town, the state's 35 and over champ is Buster Brown. He also runs a first-class racquet facility, and his comment on the upcoming competition between the two men says a lot about both of them and the sport they love.

"Tennis is a great sport for all ages. It's a terrific 'brain game.' It keeps you moving, and the health benefits are numerous. Mr. Suttle and Mr. Carey represent the best things about amateur tennis — and they're having a lot of fun, too!"

Who will win when they meet? "That's easy," says Brown, "amateur tennis!"

Winter 1992

Everett Case

The familiar face appeared on the tube during one of the NBA Championship playoff games, and a friend of mine asked, "You know who made Dean Smith possible, don't you?"

He had to be referring not only to the Carolina cagers' coach but to the supremacy of the sport around here in general. "Sure, sure," I said, "Everett Case of State, right?"

Well, my pal 'lowed as to how I was right. The man, Ev Case. He won 376, lost 133 in 18 seasons at Raleigh, 1946 to 1964. And as much as anyone, he was the catalyst who made it possible for our state to become the hardwood hotbed it is today. He made others in The Big Four rise to a higher level of performance.

Case embarrassed them into it with a string of humiliating defeats. Frank McGuire, Smith's old boss, admitted he was brought in to end such treatment. At one point, Case thrashed the proud Tarheels on 15 straight occasions.

Case's assistant, Vic Bubas, was lured away to rehabilitate Duke. Bones McKinney was given a similar charge at Wake.

Hoosier Case began play with an influx of Indiana talent. He popularized the fast break, brought in bench strength to grind you down. And he premiered the run-and-gun offensive style.

Everett Case, who died in 1966, was an old-fashioned double-breasted suit, wing tip shoe disciplinarian, according to those who knew him. One who earned a modest $15,000 a year while winning 10 ACC or Southern Conference titles.

This, at a time when most folks were still pretty much football crazy in our state. In Case's early years mighty Choo Choo Justice bestrode the gridiron. And there were football coaches around like Wallace Wade, Carl Snavely, and Peahead Walker.

Then, curiously, things began to turn 'round. Tarheel kids were shooting hoops in the back yard. National championships began coming to the state.

At Chapel Hill, a playpen like "The Dean Dome" sprouted. And on walked a wisecracking, loquacious genius, Jim Valvano. Not to mention a computer-like model of efficiency, Mike Kryzewski.

But this wiry little Hoosier, Case, why, the man never even played the game in high school. But, Lordy, could he teach it and did the fans love him! The players, they were devoted to him.

Case made a point of remembering them individually in his last will and testament.

He left all of us something, as well. A deep love of the game of basketball.

Winter 1991

Gil Coan

He lives in Brevard, does this hero of mine. Has for years now, although he was born over in Monroe, North Carolina.

My friend Nath Thompson tells me that Gil Coan farms now, after operating a successful insurance firm in their town. Tells me, too, that Gil comes to town each weekday wearing what we grew up calling "overhauls." Goes by the post office to pick up the mail and "howdys" with his neighbors awhile.

On one day there's a letter from Junior Wooten, his teammate with the old Washington Senators. On another a request for an autograph from a little boy in Canada for Gil's autograph. Or correspondence related to his commitments to the community . . . to the hospital, the United Way, the Methodist church. To Brevard College, where he played baseball.

Any of the fame that came from Coan's eleven-year career in the big leagues has been eclipsed by the respect and affection accorded him around here.

Now why is that? Why does one of our heroes end up a boozy, old out-of-focus character laughed at — and pitied — by his neighbors. While another puts their trophies down, becomes a devoted family man, a valued neighbor and friend.

What about the role luck often plays? Some of those who work a good game plan — read that life's plan — would tell us that luck is nothing more than the residue of design. And by the word design they mean devising a good life's plan and working that plan well.

The old outfielder, Gil Coan — who had several seasons batting at or about .300 — developed a life's plan, and he worked it.

The other day in Brevard, I asked Gil about one of the oldtimers. He said he remembered the fellow, but said the man was "playing his way back down" by the time Gil got to the Senators in '46. Some of those fellows just kept playing out the string.

Kind of reminds me of the old country song: "Look me up on your way down. . . . As you tumble to the ground, look me up on your way down."

Last week it was sad to see Hall of Famer Jim Palmer shelled for about five runs in a single inning in a Grapefruit League comeback try. Palmer, bless him, accepted time's verdict. He was through, and now he knew it.

The man from Brevard, Mr. Coan, loved the game and he played it well. Then he hung 'em up and went about the Game of Life. And down his way, they'll tell you he is a hero for all seasons.

As a farmer, Coan can't wait for planting time. Then the harvest. And the next planting and the one after that one. With him, there's a season for playing ball, and there's a season for living a life. Living it with and for the people around these parts.

Now, you know, that's living life like a real hero. And that's Mr. Gil Coan of Brevard.

Winter 1991

James M. (Nemo) Coleman

The Southern Railway locomotive chugged its way from the fictional Western Carolina city of "Altamont" to "Pulpit Hill." Site of the state university. The year was 1916.

The gawky young mountain boy was all arms and legs; 6'3" and something less than 130 pounds, he was. He had yet to turn 16. The two wealthy men's sons, resplendent in straw boaters and flannel jackets, had been educated at Eastern preparatory schools. Together, they contrived on the trip to make miserable the life of "Eugene Gant."

Taunts, jibes, and assorted putdowns about the boy's humble background rained down on Gant's sensitive hide. The shy country lad, however, turned the tables, using his extensive vocabulary and agile wit to counter with rapier-like retorts.

Enter the captain of the university's football squad, "Zeno Cochran," to interpose his considerable presence. "That's right! That's right! Talk to them, Gene," said a tall lad of twenty, slender and powerful, with the grace of a running horse.

This was the Saturday's hero who had "punted against the wind for eighty yards in the Yale Bowl," and he hailed from the freshman's hometown. Soon after the incident, the grid star was off to the war in Europe, where he was recognized for his bravery in the Meuse-Argonne offensive. Then "Zeno" would return to the university for the season of 1919 to lead his teammates to a state championship.

The awkward freshman in truth was Tarheel author Thomas Wolfe and the account is from his much-acclaimed novel, *Look Homeward, Angel.* In those pages, the author would describe the gallant gridder as "a handsome fellow, soft-spoken and kindly, with the fearless gentleness of the athlete."

"Fearless gentleness?" Yes, that was a trait of character one would associate with the late James M. (Nemo) Coleman, who died in his hometown in 1983 at age 88. People here tell me Coleman's personality was one part courage, one part gentleness.

Nemo, called "Zeno" in Wolfe's novel, was not only a hero to the future author that day, but to the youths he coached in later years at Bingham Military School and Asheville School in this mountaintop paradise. (Bingham even scheduled the colleges — Carolina, Davidson, Wofford; "The bigger they are. . . .")

Then the hero moved on to real estate, making his mark there, as well. He headed important civic agencies, and served on FDR's War Production Board during WW II.

An elder in his church, First Presbyterian, he served also as a Sunday school teacher. The historic sanctuary and the sermons of the spellbinding minister there also are spoken of in Wolfe's novels.

There was a time, I think, when the team captain at almost any school was a leader, not only on the field or the court, but in virtually all venues of school life. And often the captain's best qualities included the kind of "fearless gentleness" the author ascribed to Nemo Coleman. The kind of sense of what is right, what is wrong that prompted this captain to befriend the beleaguered freshman.

It was like Thomas Wolfe to show his admiration for Mr. Coleman in the way he did best. By chronicling that act of generosity and support in one of the nation's most-celebrated works of literature.

An interesting aspect, by the way, of the relationship

between Coleman and Wolfe is that it continued on a cordial basis over the years. "Captain Nemo," friends say, was not one who took umbrage to what some viewed as affronts to the local gentry appearing in the writer's works.

In fact, Baxter Taylor, one of the coach's players at Asheville School, says that Coleman related that he was asked by the author to do him a very special favor. The favor consisted of Coleman's delivering groceries and booze — "lots of booze," according to Taylor — to the Oteen cabin Wolfe occupied the last summer of his life. (Thomas Clayton Wolfe died on September 15, 1938.)

The Asheville writer was still apprehensive about the kind of reception the townspeople might accord him.

Yes, those were times, back then and a little later, when we actually summoned up images of King Arthur and the Knights of the Round Table, as we played the game. A time when my own coach would tell me, after a late hit on the rival's all-star quarterback, "Son, that's not the way we play the game at Darlington." (My eyes avoided Coach's for the rest of the season.)

Fair was fair. You didn't hit a man when he's down. And you didn't let bullies beat up on a frightened young boy of almost 16 just trying to make his way at the state university.

Spring 1992

Above, Nemo Coleman and classmates en route to UNC-Chapel Hill circa 1913, complete with celluloid collars and shoe-box lunches. Left to right: Edwin "Floppy" Hartshorn, Nemo, Allen "Babe" Williamson and Raby Tennent. Below, Coach Nemo with his boys at Asheville School in early '30s. Baxter Taylor at right end of front row.

Jim Donnan and Jimmy Satterfield

The media guides are out now, the pre-season polls are being released, and area colleges are showcasing their head football coaches at events like the Southern Conference's "rouser." It's a cordial gathering this one, with 60 or so media types attending and even organizing a softball team to defeat the coaches team 14-6.

It's also a time for golf, tennis, and good-natured exchanges among those in the coaching fraternity. "Yeah, we finally found a way to cope with the cuts in the team travel budget; I plan to load 'em onto a flatbed truck and put 'em up at the KOA Kampground before that game at your place," one of the coaches kids another.

The Citadel's Charlie Taaffe beams when a columnist mentions last season's big win over South Carolina and, over another span, the two straight over the Naval Academy. A few feet away, Marshall's Jim Donnan explains how, when he was a young Oklahoma assistant at age 30, the top job at Chapel Hill somehow eluded his grasp. After a turn-around 6-5 campaign at the West Virginia school last year — and being picked by fellow coaches to take the conference title this go-'round — Donnan looks absolutely hungry for success in his new job.

The Division 1-AA conference has eight members and they draw crowds with as few as 4,000 in attendance at East Tennessee to about 20,000 for some contests at Greenville's Furman University. By contrast, the big boys at Clemson routinely pull in around 80,000 to a home game.

But Clemson, Oklahoma, or Notre Dame this ain't. The best 1-AA loop in the country, it is. The SC often has two or more teams playing off for the national championship in that division.

Furman has gone to the "world series" of this level of competition seven times in the last nine years. And head coach Jimmy Satterfield has a record actually superior to that of predecessor Dick Sheridan, who moved up to NC State and the big time.

Also, one gets the distinct impression that Satterfield has a calculated strategy for fending off ambitious Marshall, Donnan, and his "Thundering Herd." Donnan's charges move from a 17,000-capacity layout this season to a playpen seating almost 30,000. One with 20 "skyboxes," those expensive and lavishly-appointed suites in which well-heeled corporate supporters ply customers and friends with food, drink, and football.

The personable Marshall boss discovers during the course of the rouser that a rival coach is out to administer him an awful beating on the golf course. The ever-competitive Donnan responds by rallying to win in a big way.

This Western North Carolina native *IS* hungry. And it would seem that his new employer has consciously elected to move up in the ranks of college football's elite.

Nothing really wrong with that, I guess. But it'll be interesting to see what the Paladins' Satterfield comes up with in the way of a game plan to counter that threat.

And, whoa, what a donnybrook it'll be when I-AA's national champions — Georgia Southern — begin play in the Southern Conference in 1993!

So, despite all that happy "how-the-heck-are-ya" stuff at the rouser, you can still just reach out and grab off a piece of something one pundit calls "the phenomenon of rising expectations."

Summer 1991

Jim Donnan (above) and Jimmy Satterfield (below) are outstanding young football coaches at Marshall and Furman, respectively. Their success has helped make the Southern Conference the premier NCAA Division IAA loop in the nation.

S.G. Gabriel

Shirley (S.G.) Gabriel of Appalachian State, from a famous family of athletic brothers, was selected as one of a daily's "Flaming Five" top sports stars in 1953 for the Carolinas.

Sports fans really groove on those trivia questions, don't we?

Here's one of my favorites: Name members of the NFL's all-alphabet backfield. Of course, O.J. Simpson comes to mind. Then Y.A. Tittle, the old quarterback.

From there it gets a little tougher. How about L.G. (Long Gone) DuPre of the Colts?

Now, that fourth and most elusive back. Ah, J.C. Caroline of Illinois and the Chicago Bears!

In 1953, the Orangeburg running back became one of "The Flaming Five" selected by *The Charlotte Observer*. They were the best athletes in the Carolinas for that year.

J.C. Caroline was one of just two football players picked that year. The other choice also had an alphabet monicker. His name was Shirley G. Gabriel . . . or S.G. . . . and he was a 5 foot 10 inch, 165-pound back from Appalachian State hailing from Lincolnton, North Carolina. It's "S.G." now in Rutherfordton, where he is personnel director for Stonecutter Mills Corporation.

Says Gabriel, "Yep, I took a lot of kidding back then about the name Shirley — you know, 'A football player just shouldn't have a girl's name, he can't be any good!' "

Well, S.G. Gabriel COULD play that game. He not only was a "Flaming Five" selection, he won all-state college first team honors. And Little All-America mention.

Gabriel managed to tally 111 points, second highest total in North Carolina history.

He ran for over 1,000 yards, while logging 55 minutes a game. He was even pressed into service to boot extra points and field goals.

Gabriel also was president of the App State student body. He was "Who's Who in American Colleges and Universities."

And guess what? S.G. Gabriel is still a leader. In his community. As a matter of fact, he probably is the best-known and best-liked man in Rutherfordton.

Nowadays, this kind and caring man grows his own roses and delivers them to ailing friends and neighbors. Or to the unsung heroes in that hometown.

This I know from his many friends, some of whom tell about catching S.G. doing good. People like Reverend Ken Lyon and sports trivia whiz and insurance exec Tommy Hardin.

Another friend is Dr. Dillard Morrow, his golfing pal, who asks me, "You're gonna do a hero piece on S.G.? Why in the world would you do that? Can't be because of his golf game!"

I've got an idea that one of these days, his alma mater is going to recognize Gabriel by planting his roses around the circumference of historic Kidd Brewer stadium. He and brother Steve already are in the school's hall of fame.

Another brother, Charlie, played at West Point at the urging of the late Herman Hickman, joining back-to-back Heisman Trophy Winners, Doc Blanchard and Glenn Davis. General Charlie Gabriel would become chief of staff of The United States Air Force.

But S.G. Gabriel stayed close to home. And aren't his neighbors proud he did?

Winter 1991

Cally Gault

Cally Gault remained as head football coach at the Palmetto State's Presbyterian College as his counterparts at the state's other institutions came and went. Upstate born and bred, Gault is one of South Carolina's most beloved sports legends.

"No, he was the wingback," says my brother, recalling a high school football game 48 years ago. "Tells me he was the tailback," I counter.

The player in question was one Cally Gault, and in point of fact in 1943 he alternated at both positions with a boy named Cowboy Simpson at Greenville High School. The same pattern was followed by Clemson a few years later with a pair of All-Americans, Bobby Gage and Ray Matthews, doing the swapping off.

The game with Greenville is one my brother would rather forget. After polishing off the defending state champions, Anderson High and Gage, the previous Friday, our boys journeyed to the big city of Greenville to face off with the Red Raiders at mammoth (to us) Sirrine Stadium.

"Oh, I remember your brother, the tailback," said Cally Gault, "and we thought y'all were gonna beat us! The year before that Anderson crowd had whipped us good."

Cally and his Raiders took us handily, 65-0. But at season's end, four of our players — including my brother — were awarded grants-in-aid by Clemson's Frank Howard. The Baron of Barlow Bend, on the other hand, told GHS coach Speedy Speer that Gault was just too small to rate one.

Cally instead went to little Presbyterian College to play his ball — and in time simply became a living legend at the Clinton, South Carolina, school as its all-time winningest coach. Since 1984 he has been its immensely popular athletic director.

A recent inductee into the NAIA hall of fame, Cally also is a member of the State of South Carolina Athletic Hall of Fame. His 22-year tenure as PC's head football coach was longer at the time of his retirement than all of the state's college coaches' combined. He is the only football coach in the Palmetto State ever to win five Coach of the Year awards.

Before answering his alma mater's call, he became a successful high school coach at North Augusta, winning three state championships while registering a four-year undefeated string of 42 consecutive games. That's a mark that still stands.

"As a coach, Cally was the greatest motivator I ever saw," says Bob Calhoun, a six-foot-three, 190-pound tackle who's now a dentist in Asheville. Calhoun continues, "I always thought he would move up to the big time."

Why would the old coach do a thing like that? Talking to him at PC's homecoming game with Wofford, I think I discovered why he chose not to move on.

Oh, call it small-time football at Presbyterian and you'd be right. Pre-teen boys play a pick-up game in the end zone as a crowd of 4100 gathers. Gault is a man, though, who loves everything about this institution with just 1200 students. He's had a good life here, and he in turn has been good for the game. Same thing as well for the school he has served with such dedication for over three decades.

"Do I miss being out there?" he asks. "Only on game days and only when the coach leads the team onto the field."

He adds quickly, to the accompaniment of vigorous nods from Ms. Gault, "But I sure don't miss waking up at four in the morning on Sundays, haunted by something wrong with our defensive alignment." It was at those times that the coach would rise and set out for the gym to view game films for 10 to 12 hours.

"I knew when to leave," says Cally staring straight ahead at the action on the field. His left hand grabs my bicep as he talks, wincing slightly at a broken play on the natural grass below. He relaxes then and smiles as winless PC goes out at halftime with archrival Wofford. They're tied at 24.

"I remember losing it that last season after a bad call at Newberry and blasting the officials," the coach says. "You didn't pull a Woody Hayes and slug somebody, did you Cally?" I asked. No he didn't, but he didn't like losing his self-control, either. So like Bo Schembechler at Michigan, Gault made his decision to leave while he was ahead. The athletic community who knew both were surprised.

About Schembechler, Cally reminds us, "He started his coaching career right down there on that field. He was just out of Miami University in Ohio."

"Yep, we have 100 boys out for the team and 68 get some kind of scholarship help, but there are just 25 full scholarships to split up," he points out. That's about the way it was when I attended the school for a year in 1950-1951. (Fact is, I got a one-fourth scholarship for academic standing and another one-fourth for being a Presbyterian, which was pretty outstanding since I was a Baptist at the time.)

Then I realize it has been 40 years since I have been to a game at the old school. Too long, but it wasn't the same school it is now. Cally Gault wasn't there yet coaching.

Back in the press box now, the public address announcer gives the South Carolina-East Tennessee score. Tied, did he say that? Then a Greenville reporter, "Hey, just wait for the big one, the real big one is down at Sanford today."

He is referring to the Clemson-Georgia contest later that day at the Bulldogs' historic Sanford Stadium.

Leaving the press box to greet old friends in the stands then, I am compelled to comment to the reporter from Greenville, "No, no, this is the real big time, Bubba." Any time a school has a class act like Cally Gault around you are big time — whatever may be the attendance, the number of scholarships, or even the final score out there on Bailey Field.

Fall 1991

The Four Horsemen and Chuck Collins

Chuck Collins was one of several Four Horsemen-Era Notre Dame stars to coach college elevens in the Southland. Collins ranks fourth on UNC's list for career wins.

It was a memorable phrase coined by an East Tennessean, one who talked like us. Who thought like us.

To him, heroes were epic in their deeds. And yet as emotional as his prose could be (he also wrote some poetry), he could make language march at his very command with discipline and precision toward a given objective.

His name was Grantland Rice — "Grannie" to his many friends — and in days gone by, coaches and boosters would preach at us using Rice's lines, "When the Great Scorer comes to write against your name/ He writes not whether you won or lost/ But how you played the game." We revised that to read, "He writes not whether you won or lost/ But how you wore your uniform."

And so it was that a winsome cheerleader named Tootsie lifted me sky high when she expressed admiration for, well, the way I was "poured into them game pants." She only could have noticed during pre-game warmups, for I was on the bench for at least 57 minutes out of every 60-minute game.

Well, none of today's sportwriters write with the same inspiration Rice did. And yet no one can resist this one. It is the very essence of what college football was — and may still be: truth, poetry, folklore . . . and a little reality.

Here's what I mean. Sixty-seven years ago Rice wrote himself a place in sports history. He penned these immortal lines: "Outlined against a blue-gray October sky, The Four Horsemen rode again. In dramatic lore they are known as Famine, Pestilence, Destruction, and Death. Their real names are Stuhldreher, Miller, Crowley, and Layden. Together they form the scythe of the South Bend cyclone."

We can envy Notre Dame football, its traditions and its broad national appeal. We can exult when one of our Southern teams beats them head-to-head for a national championship. But their majestic presence endures, and we watch what they do so as to go and do likewise.

Oh, about those Horsemen. Two of 'em came South after playing days. Several more of The Irish followed in their footsteps, just as several others had preceded them. We hired them as coaches in the 1920's and 1930's to help us achieve national supremacy in the sport — and perhaps to evoke memories of "The Lost Cause." You know, as in "The War of Northern Aggression."

The Great Rockne dispatched halfback Don Miller to coach at Georgia Tech. Miller completed law school at Atlanta's Emory University, to boot, and returned home to practice in Cleveland. "Sleepy Jim" Crowley joined another of that 1924 team, Harry Mehre, then head coach at Georgia. He spent three seasons in Athens, 1927-1929. Another son of Notre Dame eventually became head coach at Alabama. His name was Frank Thomas.

Yet another from South Bend, Jack Meagher, took over at Auburn. And a neighbor of mine in Columbus, Ohio — All-American guard Jack Cannon — journeyed this way to join Meagher's staff. Earlier, Frank Shaughnessy had Clemson for a season, and a bit later Rex Enright, an Irish fullback, took over the top job at South Carolina. Add to that long list the names of Chet Wynne at Kentucky and Tom Lieb at Florida.

In our own state one of the "Seven Mules" who cleared the way for The Horsemen landed at Chapel Hill as head coach of the Tarheels. He was Chuck Collins, who in an eight-year stint at UNC (1926-1933) posted a mark of 38 wins-31 losses-nine ties. The mark puts him in a tie with Ray (Bear) Wolfe for fourth place on the all-time wins list behind Dick Crum, Bill Dooley, and Carl Snavely. In Collins' second season, 1927, the 'Heels dedicated Kenan Stadium, still one of the nation's most beautiful college layouts.

Each of these Horsemen went on to achieve good things after their playing days. Harry Stuhldreher became head man at Wisconsin, as did Crowley at Fordham. Miller became a respected attorney and civic leader in Cleveland. And Elmer Layden was named Commissioner of the National Football League, serving also as top man at the South Bend campus on his way up.

It was Layden who was the heavyweight of the four at about 170. Together they averaged less than 165. But Grannie Rice's "Four Horsemen" inspiration led them to attain a level of prominence — and palpable fame — that lives on today in the hearts and minds of those of us who show up on Saturday's at stadiums across America.

It is we who experience the thrill that comes when our favorite team, like The Four Horsemen, "Shake down the thunder from the skies." And, to think, it was one from our own native ground who uttered the phrase that created all the magic.

Fall 1991

The Four Horsemen: HB Don Miller, FB Elmer Layden, HB Jim Crowley, QB Harry Stuhldreher of University of Notre Dame.

Ed Hamilton

We called him "Irish Eddie," and he didn't make the team.

Edgar B. Hamilton is a respected and dignified banker in Shelby, North Carolina, these days. But back in the early 1950's, he was a skinny but promising Georgia Tech quarterback. A passer he was, with a slingshot for an arm.

Yep, a great one at Atlanta's North Fulton High School. And the brother of my University of Georgia pal, Joe Hamilton. The younger sibling was the poised and self-confident field general of the Yellow Jacket freshman team.

Pre-ordained to decide Ed Hamilton's

Irish Eddie Hamilton of Georgia Tech gained his distinctive nickname after tossing four TD passes in a Turkey Day contest with arch rival Georgia. His leadership, however, has been even more noteworthy as a Shelby banker and major player in civic endeavors.

fate as a varsity performer, however, was the kindly head coach at the school, Bobby Dodd. The name, you'll notice, rhymes with "God."

(One-time jacket tackle Rees Phenix, a roommate in my post-graduation days in the Georgia capitol, told us he just had to sign with Tech out of high school because Dodd reminded him so of his childhood idol, cowboy star Roy Rogers.)

Well, the coach decided he wasn't going to be too keen on my friend "Irish Eddie." (I'll explain that nickname later.)

"Tell me, son, what's your name?" Dodd asked Hamilton at an early season scrimmage. Adapting a line from the fabled George Gipp in the movie of the life of Notre Dame's Knute Rockne, the brash signal caller spoke up, "Why, Hamilton of Tech, sir!"

The Roy Rogers smile quickly faded from the coach's friendly visage. He plainly was not amused.

Arkansas' Frank Broyles was Tech's offensive boss then, and Irish Eddie soon became Broyles' favorite for the starter's role. But rival Bill Brigman was the favorite of God, er, Dodd.

In that freshman season, the frosh fell behind the Bullpups of Georgia, 20-0. A desperate freshman coach barked out, "Hamilton, get in there and throw that blamed thang!"

Throw Eddie did. For four touchdowns and a 38-26 win before almost 40,000 at Grant Field. It was Hamilton's finest hour on the gridiron.

By spring practice time, Ed was just coming off a lengthy bout with the flu. Or so he and the doc thought. He had mononucleosis, and the slingshot didn't even know it.

Hamilton was dragging about listlessly in practice sessions and kindly Coach Bobby Dodd mistakenly labeled him a loafer. And that was the end of that career.

He got over it, I think. Concentrated on his books and was a student leader. And in time, Hamilton met and married former Governor Dan Moore's daughter, Edith. Moving to Shelby, Irish Eddie got on with the serious business of becoming a banker, a husband, and a father.

They'll tell you over in Shelby that the old hero made the team in that town. I understand with almost everything good in that loveliest of villages, Ed Hamilton has had a hand in making it happen.

Now, why do I keep calling this fellow "Irish Eddie?" That's because as a student – and refugee from the Tech varsity – he journeyed over to visit brother Joe and attend one of our home games in Athens.

The whole time, Hamilton kept tuned to a radio to get the Tech-Notre Dame score. He kept predicting a great victory for The Ramblin' Wreck.

Uh, uh: 27-13 for the guys from South Bend, The Irish. The final score in, John "Kid" Terrell just crowed out loud, dubbing Hamilton, "IRISH Eddie!"

I saw Ed right after Thanksgiving last fall, and he insisted on taking me to lunch. His lip is not so quick now as it was with kindly Coach Bobby Dodd.

The food arrived and Ed asked, "Mike, can we bow our heads now to thank Him for all we have to be grateful for?"

Ask the blessing we did, and I got to thank the Lord for this bona fide All-Shelby selection there with me. His patient and gentle ways have brought us far more meaningful victories than the one he took home from Grant Field that Thanksgiving Day game four decades ago.

Summer 1991

Joe Harand

High-scoring Vic Bubas was NC State's shotmaker. The other guard was the playmaker, Joe Harand. That was over a sensational four-year span, 1946-1950, for state's incomparable Coach Everett Case.

Those years were the "golden age" for college basketball in the Carolinas, according to many students of the game. No fewer than three future head coaches were active then as players: Horace "Bones" McKinney (later head coach at Wake Forest), as well as Case's Norm Sloan (who ended up as head man at Florida) and Vic Bubas (to be top dog at Duke).

However it was the Tenafly, New Jersey, transplant — Harand — who himself achieved a kind of notoriety few of his neighbors in Shelby know about. A seen-but-not-heard type of competitor, Harand once bore the brunt of an unintentional slur — but nevertheless, a slur — perpetrated by the Carolina cage mentor, Tom Scott. ("Actually," says the Shelby man now, "he was a very fine person.")

Vowing prior to the contest his 'Heels would be all over the high-scoring Bubas, Scott said he wouldn't worry about the other guard, Harand. "Can't shoot, you know," was the Carolina man's rap.

"Didn't bother me," says the old playmaker, "I'd heard it before." But the irascible Case had plans for exacting a measure of revenge on behalf of the unselfish team-playing guard.

Flashy Sam Ranzino was switched to Harand's guard slot, as the 6'4" Harand moved to Ranzino's vacated forward position. And the high-scorer that night of the big grudge match?

None other but Mr. Joseph Harand. Justice was done, Case taught Carolina a lesson, and State was on its way to another Southern Conference championship — and its fourth straight NCAA bid. (Case, it's worth noting, whipped Carolina 15 straight times.)

But now for the rest of the story about a coach's admiration for a selfless player. The Shelby resident was a serious-minded engineering student and had to attend labs on weekday afternoons. So the coach called the team together, "Mr. Harand here needs to make his labs, so henceforth we will practice at 7:00 p.m." So they did, and Harand became the school's first basketball player to obtain a degree in the dyeing discipline of State's tough textile engineering school.

That diploma — as well as native Germanic determination and hard work — have stood the old playmaker in good stead. At age 65, Harand has had a solidly successful career, heading his own company in the Cleveland County town.

Coach Everett Case always remembered Harand for his dedication and effort. And upon Case's death, the team-player was among those remembered in the stoic old Hoosier's last will and testament.

A while back that remarkable Wolfpack team celebrated an anniversary, something that I gather they do with considerable regularity. "Where's Paul Horvath?" someone shouted out. Turns out the old center was on television's "Wheel of Fortune," winning $46,000.

Well, that wheel of fortune already had spun Joe Harand's way on a night 41 years ago in Chapel Hill when he emerged from the shadows to win The Big One with his high scoring.

Folks tell me he's been that kind of citizen around these

parts too, scoring high with friends and neighbors alike. In things like loyalty, unselfishness, and solid performance. Coach Everett Case, they'll remind you around here, knew real character in a young man when he saw it.

Fall 1991

Joe Harand (L), was an Everett Case favorite at NC State because of his intelligence and team play. Harand once had his day in the sun as Case taught Carolina cagers a memorable lesson. (Vic Bubas at right.)

Dave Hart

Dave Hart, a commissioner who took the Southern Conference to new heights, was a well-traveled coach and administrator who never met a stranger. Son Dave Hart, Jr., carries on the tradition, leading East Carolina University into the ranks of the nation's top athletic programs.

To me, he has always seemed the extroverted, hail-fellow Irishman. One who just happened to be at the top in athletic administration and not, as chance might have it, politics or entertainment.

It's hard not to like Dave Hart, recently retired commissioner of the Southern Conference and an individual adopted some six weeks after being placed in a Pennsylvania orphanage.

With a tremendous boost from his adoptive Catholic parents, "Davey" Hart – as he was known in his playing days at the University of Georgia

— has grappled his way to success and acceptance in one arena after another. In May the irrepressible Hart was named to the college athletic directors hall of fame.

Coming back from 30 missions as an Air Corps pilot in World War II, having won The Distinguished Flying Cross and The Air Medal with Oak Leaf Clusters, he returned to Georgia and then enrolled at a Catholic institution in his native Pennsylvania, St. Vincent's College.

At that state's Connellsville High School Hart had been "the other halfback" to Notre Dame's Heisman Trophy-Winning Johnny Lujack. And at St. Vincent's he was the team captain and led it to a 7-6 Tangerine Bowl victory over Emory & Henry.

After graduation came positions in both high school and college coaching. As an assistant at Kentucky and Navy, then as head coach at the University of Pittsburgh. Leaving the coaching ranks, Hart shifted to Robert Morris College in Pittsburgh, next becoming athletic director at Louisville. But it was in 1978 that he came into his own as athletic director of a highly acclaimed program at the University of Missouri.

Sportscaster Bob Prince in Pittsburgh called him "Bullfrog," saying that Hart was "always leaping from one lily pad to another in search of a better job."

As a high-spirited, high-energy A.D., Hart became especially adept at fund raising. I remember how before a group of AD's from another conference — and speaking on something he virtually invented, sports marketing — he remarked, "I'm Catholic, you know, and we can flat raise the money. We pass the collection plate four times, then we lock the door and search 'em!"

A front-page *Wall Street Journal* story was headlined: "UNIVERSITY OF MISSOURI SCORES VICTORIES AT GATE AND ON FIELD — ATHLETIC DIRECTOR'S HUSTLING KEEPS PROGRAM IN BLACK AS MOST SCHOOLS FLOUNDER." His was a clean program to boot.

In 1986 the Southern Conference landed Dave Hart as commissioner and everything came up roses for the league. (Its membership includes Furman, Appalachian State, Marshall, and Western Carolina; VMI, East Tennessee, The Citadel, and UT-Chattanooga.) The Southern was struggling before Hart's arrival on the scene, but it's acknowledged today to be preeminent in the 1-AA classification of the NCAA.

In Dave's five-year term as boss the conference has enjoyed ever-increasing sponsorship dollars and has begun to share revenues with its member schools. Georgia Southern, I-AA's reigning national champs of the gridiron, is joining the league. Davidson has returned, bringing membership up to 10 institutions.

In addition, the conference's basketball tournament has experienced great success in its new location, Asheville, the loop's headquarters city for five years now. The ESPN Network comes to televise the event nationally each year.

Dave Hart's legacy? He brought modern marketing techniques to bear on problem areas and put The SC on a sound financial basis.

The Pennsylvania transplant who first came South in 1943 intends to live out his days as a Tarheel. That's not such an uncommon occurrence, you say — an Easterner won over by the Southern way of life?

But how about this now? Dave Hart Jr., athletic director at Eastern Carolina, is said to be the favorite to bag the same job with perennial 1-A power, Pitt. Question is, after putting in his time in Steel City, will young Dave return to serve the cause of collegiate athletics down our way? Some Eastern sportscaster might label him "Bullfrog Jr."

So what? Willingness to move as you move up, that's often the route to the top in college sports administration nowadays. As the senior frog once wrote in a credo for players at Mizzou: "A winner is ready to tackle something new. . . . A loser is prone to say it can't be done." And in the same declaration of faith: "A winner is challenged by a new problem. . . . A loser doesn't want to face it."

So, if someday Dave Hart Jr. comes home again, our area may be the winner with the junior member of the clan as it has been with "The Old Bullfrog" himself.

Summer 1991

Gomer Hodge

Harold Morris Hodge picked up a nickname, "Gomer," that has stuck with him for almost 30 years. He answers to "Gomer" readily, bowing to the inevitability of it all after so long a time.

Beginning in 1963 and playing his way up from a Class A Dubuque, Iowa farm team, the Rutherfordton native eventually enjoyed a season in the sun. That was in 1971 with the American League's Cleveland Indians, and he was at the time a mature 27 years of age. The tribe that year was going nowhere under managers Alvin Dark and Johnny Lipon. They finished dead last in the AL East.

Sports Illustrated "discovered" Hodge in that single campaign in The Bigs.

Gomer Hodge was a Cleveland Indians favorite, whose loyal fans seated themselves in a special section they named "Hodge Lodge." His friendly, down home manner endeared him to legions of supposedly cold-hearted fans North of the Mason-Dixon Line.

Gomer Hodge had always endured a lot of good-natured ribbing from teammates. And someone had labeled him "Gomer," as in Gomer Pyle of network television fame.

"It was because of my uncle's Southern dialect — somewhat like Gomer Pyle's — rather than his appearance," says Mrs. Walter Dalton, a one-time Shelby resident now residing again in her native Rutherfordton. If you remember Pyle, played by actor Jim Nabors, he was a gawky, likable Southern bumpkin.

Well, this Gomer became an instant fan favorite. In his first big league game, he merely hit a single, a double, and a triple — and, yep, a home run!

Sports Illustrated, or maybe it was another outlet, asked, "You do realize you're hitting 1,000 now, don't you?" "Goll-eee, seems like it ought to be 4,000 to me!" That was what Hodge was supposed to have said.

The sports weekly's piece also dwelt upon "Hodge Lodge," the bleacher section his adoring fans staked out as their own and named in honor of the popular ballplayer. (At various times, Hodge played first, second, and third base.) About the same time, I believe, the Seattle fans staked out "Perry's Perch" in honor of Gaffney's Hall of Famer, Gaylord Perry.

Well, forget the lifetime B.A. of .205 and don't assume his career ended there at Cleveland, the city they call "The Mistake on the Lake." Lodge has toiled for 20 years now in the minors — and high minors — as a batting instructor, coach, and manager. He currently is AAA batting instructor for the American Association's Indianapolis Expos, the prime farm property of the National League's Montreal entry.

You know, some of us carry around nicknames like an albatross for years. We don't much like these monickers that were laid on us, but we answer to them. I asked Hodge if he perhaps had a problem like that.

Well, as a matter of fact . . . I know, I lived up North for almost 20 years, and I had it up to here with negative stereotypes of Southerners. You talk slow and in their minds, you're a Gomer Pyle.

You graduated from a Southern college they never heard of, so you must be mentally defective. You're a 25-year-old single man living in Philadelphia – and you like girls. So you, my friend, are a Rhett Butler. (Now, come to think about it, that wasn't all bad!)

So it was in Baltimore one night, someone seriously suggested, "Gomer, tonight when you get a single, stumble over the bag and fall flat on your face just like Gomer Pyle would." That got Hodge hot under the collar, and could you blame him?

I asked him if this kind of stereotyping had affected his career progression, and Gomer 'lowed as to how it might have in some ways. The idea was that maybe management had on occasion taken all his hard work and dedicated service for granted. That they were thinking, "Good ole Gomer, just give him the short end of the stick; he's such an easygoing old boy, he'll take it."

That aside, Gomer Hodge has had a good life. He married late, has two young children, and people like him a lot.

They should. He's one of us.

Fall 1991

Terry Holland and Bob McKillop

"Gable's back and Garson's got him!" That's what Hollywood proclaimed as it ballyhooed Clark Gable's return from what some have called "The Good War," World War II. The incomparable Greer Garson was the leading lady in that first post-war movie, whatever its name was.

A few miles north of Charlotte there's a venerable college, Davidson by name, with some 1550 students. It's time for them too to welcome back the conquering hero, Athletic Director Terry Holland, named to the job in May 1990. So, "Terry's back and Davidson's got him!"

Holland is, along with Lefty Driesell, one of the winningest basketball coaches in the 154-year history of the school. And he was an all-time great as a Wildcat player out of Clinton, North Carolina (population 7500 — over there near Turkey and Calypso, North Carolina).

Add to those credentials, 16 seasons at the helm at the University of Virginia, where he was the school's greatest coach in memory. With the Wahoos he had 10 teams invited to the NCAA tourney and another five were extended NIT bids. Five times he was Coach of the Year.

Holland's winning percentage at UVA was a gaudy .643. In five outings at Davidson, he notched a .681 mark. Of course, those are some of the stats, but one can appreciate the man — and his qualities — in so many other important ways.

Here's what I mean: at the Southern Conference's recent tip-off, a columnist approaches Holland and Conference Commissioner Wright Waters. "Wright, what's the idea of inviting this fellow from the Food Lion television spots?" The query draws a self-conscious chuckle and a big grin from Holland, who bears a striking resemblance to C.E.O. Tom Smith of the big food chain. Waters, he just guffaws.

Terry Holland is one to enjoy a laugh — and pause to smell the flowers along life's way. "A people person," he definitely is.

That's going to help him in his relatively recent role at the smallest institution competing in the NCAA's big-time Division I. His head basketball coach there — a man much like Holland in his personality and his winning ways — is Bob McKillop. McKillop needs Terry's steadying influence if the Wildcats are to remain at that level of play and compete again in the Southern Conference beginning with the 1992-1993 season.

The school, with its strong ties with the Presbyterian denomination, detoured from the Southern, spending a while in the Big South conference. Now like Gable, they too are coming back.

What kinds of teams will Holland and McKillop field at Davidson? Well, this year's has a decided international flavor. Two of four returning starters are from overseas. They are Detlef Musch, a 6-11 center from Germany and 6-8 sophomore forward Janko Narat from Yugoslavia.

Freshman Franco Ferroni, a 6-10 forward from Italy figures to vie for playing time, as well. And, oh, starter Jason Zimmerman is a 6-2 guard from Warsaw. Indiana, that is.

This international thing, it's not out of character for a college with study programs at distinguished institutions in France, Germany, and England (Cambridge University, no less). Nor is it unexpected at an institution whose entering freshmen score an average 1230 S.A.T. That kind of standing has wide appeal for the academically well-prepared from anywhere.

Another impressive fact: Davidson has had 21 Rhodes Scholars — tops, I believe, among schools of any size in the Southeast.

With skilled and personable leaders like Bob McKillop and Terry Holland for their athletic program, the Wildcats seem intent on winning again in the big time. Too, they may start playing the role of David, the Giant Killer. They've done it before.

Now if only Terry Holland would quit doing those nutty grocery store TV spots. He and McKillop must start stealing away some blue chippers from the state's Big Four. It's something Holland's done before at Davidson.

The new recruits had best polish their foreign language skills, though. In addition to their stats, that is.

Fall 1991

Terry Holland (L) and Bob McKillop (R) are a dynamic duo for Davidson College athletics. Home-stater Holland is committed to his school's excellence in academics, and McKillop is his choice to lead the Wildcats as they return to the always-tough Southern Conference.

Frank Howard

Down the road apiece, there's a great university in the Upstate of South Carolina. Its ancient traditions and ceremonial rites are many, but none are so cherished as those associated with a kind of field of dreams known as "Death Valley."

Clemson's Frank Howard is a living legend at the school. "The Baron of Barlow Bend" tops almost any other pigskin mentor in terms of career wins and tall tales.

On crisp Fall Saturdays there, Tiger Paw symbols abound and the varsity goes into its 25-second run down the hill as the fans go absolutely berserk. And before that run to glory the players traditionally seek the blessings of the gridiron gods by rubbing something called "Howard's Rock."

The man for whom that lucky charm is named is profiled under the heading "Living Legend" in the Clemson University press guide. His name is Frank Howard, known to some as "The Baron of Barlow Bend."

In a half century at the school, the Barlow Bend, Alabama, native has seen Clemson grow from an all-military, all-male A&M institution to the 16,000 student multiversity it is today. And it was The Baron himself who took the Tigers from an 11,000 capacity stadium — and seasons of losing to the likes of Oglethorpe, Furman, and Presbyterian — all the way to top 20 rankings with memorable teams that won bids to the Sugar, Orange, Gator, and Bluebonnet bowls.

He's in The National College Football Hall of Fame. But at age 82, Coach Frank Howard is as down home, as unpretentious as ever.

A couple of weeks ago, I spoke with half of the "Frank and Peahead Show," a long-running comedy series replete with cornpone humor and devastating putdowns. The other half was the hilarious Douglas Clyde (Peahead) Walker, one-time head football coach at Wake Forest University.

"Oh, yeah," The Baron drawled, "I got plenty of stories on Peahead, but most of 'em couldn't be printed in your book."

Then he starts in. Seems Walker was playing Clemson and was identifying his coaches and players for the stadium guard at the visitors gate. At the end of the process, the guard asked "And who are you?"

"Why, I'm Coach Walker!" The guard, probably on orders from Howard, acted unconvinced and summoned Wake's captain, Pat Preston. "No, sir," said the captain, "I never saw this man in my life."

Then Howard chortles, "Peahead told me Preston sure knew old Peahead's identity" on Monday after the player was put through all those laps.

On another occasion, The Baron rose to meet the challenge. His quarterbacks fell to injuries . . . one, two, three. So he calls forth a running back, Charlie Waters, later an all-pro with the Dallas Cowboys. "Son, you heard I'm an offensive wizard, right?"

"Right, Coach!" the halfback exclaims.

Says The Baron, "Well, poof! You're my quarterback!"

My late father, a school superintendent and one-time coach, told of going to Howard's school in the 1930's and finding head coach Jess Neely and a shirtless Howard hammering out metal yard markers in the blacksmith's shop. (The two were the entire coaching staff.) Dad said Frank Howard was the finest looking physical specimen he ever saw.

"Lordy, good thing he can't see me now," quoth The Baron.

Another time Frank was the starter at the district meet; and, because no official starter's pistol could be found, Dad handed the coach his Army .45 revolver.

The Baron fires straight up, an entire tree limb and dozens of pine cones shower the runners.

"Yeah, and then I disqualified your dad's relay team for dropping the baton. Those old mountain fellows pulled so many knives on me I thought I was gonna have to use your dad's .45 to shoot my way out of there."

Those hotheaded mountaineers had their shortcomings, I guess. But I figure if coach had just told a couple of his favorite Peahead Walker stories, they'd have ended up giving him totin' privileges at some of North Georgia's finest stills.

A rare gift, humor, that's what the Lord passed on to Coach Frank Howard. And along with it, a way of convincing the little fellow he was just as important to Howard as the big money fellows in the bright orange Cadillacs with the IPTAY decals.

All of us little fellows, Coach, we appreciated that, too.

Winter 1991

January 1, 1942

Tomorrow's a day for viewing football's bowl games. That, and for many more of us, enjoying the pre-game parades. (And for my part, the best of these are the Rose and Orange extravaganzas.)

But let's go back in time now — 50 years ago to New Year's Day, 1942. "The Granddaddy of Them All," The Rose Bowl, had no such pageantry that year. In fact, the contest was not even played at its usual Pasadena, California, location, moving instead to Durham, North Carolina.

As Americans rang in the New Year following the attack at Pearl Harbor on December 7, 1941, we were grim-faced and in no mood for frivolity. A world on fire loomed ahead, and it was World War II.

Officials at Duke University (9-0 that season) stepped in to keep the Rose Bowl tradition alive. "Play the game here," they said. The West Coast, it was felt, was a prime target for Japanese bombers and warships. The war department was agreeable, and the opponents — Oregon State University — were eager to test the mettle of the high-riding Blue Devils.

Old-timers recall the day with a nostalgic sadness. The home-staters were upset, 20-16, by a team with the improbable nickname, "The Beavers."

Coach Wallace Wade and his team were ready to march off to war, The Old Master leaving Durham with an 11-year slate of 85 wins, 18 losses, and 2 ties. He had brought The Blue six conference crowns and a pair of Rose Bowl bids.

That '42 bowl game symbolically ended the era of pre-World War II college football. It was a glorious span when the single wing formation and its cousin, the Notre Dame Box, were standard. Players wore leather helmets and high-top footwear. There were few face masks and even fewer timid competitors.

Only four bowls were around then: The Rose (the first and oldest), Sugar, Orange, and Cotton. None had any specific tie with a given athletic conference. And when one of our Southern teams were tendered an invitation we all gathered 'round the old Philco to cheer them on.

The wartime brand of football that was to come was to prove as synthetic as the "Postum" coffee substitute we used to wash down that disgusting Spam we consumed. Seventeen-year-old freshmen, too young to be drafted, and 4-F's physically unfit for combat filled out rosters of schools experimenting with something called "the T formation." West Point and Annapolis commandeered most of the pre-war talent. And only a handful of programs challenged them — Notre Dame, Penn, Georgia Tech, Ohio State — and, yes, Duke.

The conflict ended, veterans returned to campuses in 1945 and 1946. The starting fullback for the '43 national champions, The Irish from South Bend, was Jim Mello. In '46 Mello found himself the fourth team fullback on the national titlists at his old school. "And that was a gift," says one of his contemporaries.

Wallace Wade came back to Duke, but it was never the same for him there. The game had passed him by, and he was 25-17- and 4 in his last five campaigns.

Even so, Duke had performed a valuable service, hosting the Rose Bowl and providing a bridge over which college football could continue during what some call "The Good War." And a half century later we ought to say, "Thanks, Duke University, we needed that."

Fall 1991

Valerie Lemoignan and Judy Rose

There's an advertising slogan instantly recognizable to most of us, one for a widely-known tobacco product. And it testifies to the fact that women indeed have come a long way.

With all this, however, I don't want to trivialize anything about how far women's athletics at the collegiate level have come in the past couple of decades. Granted, it can be argued that it took the catalyst of Federal legislation to bring about the progress.

Title IX of The Equal Opportunity Act of 1972 undoubtedly did clear the way, but women athletes – along with their coaches and athletic department administrators – certainly took things from there. More than the landmark legislation itself, these individuals deserve credit for the strides made since '72.

As it so happens, women administrators at two of our area institutions epitomize the progress toward newly-found competitive status in this field. They are Judy Rose, athletic director at UNC-Charlotte, and Valerie Lemoignan, Assistant AD and track/cross country coach at UNC-Asheville.

Together they represent the present and the future of woman's athletics at our colleges across the country. Rose, a part of the UNCC program since 1975, was named athletic director in June of this year. So chosen, she became one of just three females to hold the top job at a higher NCAA classification school across the U.S.

A native of Southampton, England, and a member of the British Olympic Team in 1984, Lemoignan is an administrator on the way up. Both personally and professionally, she would seem to have "the right stuff."

Head coach of both men's and women's cross country teams at the Asheville institution, she also heads up the women's track squad. Among her credentials: All-conference for four years and winner of seven different conference titles at Murray State in Kentucky; and NCAA finalist and Olympic competitor in the heptathlon, a grueling seven-event test of endurance and skill.

Lemoignan's coaching reputation is being established. She's been at UNCA for just four years, after a stint in grad school at the University of Houston. But it's her bent for college athletic administration that is likely to dictate her next career move, not her talent for coaching young athletes.

At Asheville, with its 3,200 students, she's NCAA compliance director and serves as admissions liaison with the university's administration. She also is one of four female coaches out of the school's compliment of 12.

For Valerie Lemoignan, the example of such administrators as Judy Rose at 14,000-student UNC-Charlotte is bound to be a positive one. Another might be that of Dave Hart, recently retired commissioner of the Southern Conference in the city of her residence. Pittsburgh broadcaster Bob Prince once conferred the nickname "Old Bullfrog" on the personable Hart — he was always "lookin' to jump from one lily pad to the next one in search of a better job," quipped Prince.

It was true, Hart admitted, adding quickly that "that's the way it works in college athletic administration." (And in college coaching, as well, we might add; just ask Oval Jaynes, the widely-traveled former Gardner-Webb coach who just landed the top job at the University of Pittsburgh.)

Tell me, do Lemoignan's and Rose's jobs sound like just so much "woman's work" to you? Well, to me it represents an ideal coming together of athletic achievement and career preparation with opportunity.

Of course, you must understand that I'm the father of a right talented 16-year-old female competitor — an only child — in the high jump, hurdles, and relay events. So I'm appreciating more and more what's happening in the world of women's athletics these days.

And that being the case, perhaps you'll forgive me — if and when my daughter wins a region title — if I turn around to those assembled and proudly announce, "That's my girl!"

Fall 1991

(Note: In June 1992 Lemoignan moved to a key managerial post with the National Collegiate Athletic Association: MC.)

Judy Rose

Valerie Lemoignan (above) of UNC-Asheville and Judy Rose (page 62) of UNC-Charlotte, are models for young women seeking to reach the top in collegiate athletic coaching and administration. They prove that it wasn't just Federal legislation that made the goal attainable.

Joseph H. McConnell

He was president of the NBC Network, Reynolds Aluminum, and Colgate Palmolive.

John F. Kennedy named him to an Ambassadorial rank post for International Satellite Television. The program grew into what we now call COMSAT.

He signed Groucho Marx to an NBC contract to appear in a comedy series. And he palled around with a former prize fighter from Cleveland named Packy East — later known as Bob Hope.

To Joseph H. McConnell, though, it may be that his proudest of achievements came as a scrappy, 165-pound center on the football team at tiny Davidson College, 1924-1926. He was named to the center position on the all-time Davidson eleven.

Mr. McConnell — you really don't call him "Joe," definitely not "Joey" — lives in Brevard, North Carolina. Has for almost 40 years now.

McConnell starred for the Wildcats' 1926 squad that whipped Carolina, Duke, and State. And tied Wake Forest at three.

He also was a standout in 1925. That year, they tore into The Big Four, as well. The Wildcats thumped Duke (then called Trinity), 26-0, and topped State 9-0. They tied Wake at seven, losing only to Carolina, 13-0.

Years ago, the poet wrote, "Give me men to match my mountains, men to match my plains; Men with empires in their purpose, and new eras in their brains." That was when the pioneers were surging westward. Nothing could stop them.

Well, Mr. Joseph McConnell is a real pioneer, in my book. Nothing stopped him, either. And the lessons he learned at Davidson, pinning back the ears of The Big Four teams . . . that probably prepared him for achievement on a grander scale in another competitive arena. An arena in which ideas are shaped, where seemingly routine decisions have a multi-million dollar relationship to the bottom line. And where the slip of a single ratings point can mean curtains for the most promising of careers.

In these situations, no kindly coach on the sidelines sends in a play. Or soothingly explains away a crushing defeat in assembly on Monday.

McConnell was an all-out competitor at Davidson in the 1920's. I understand he was much the same in the corporate boardrooms he inhabited for decades after his college days.

McConnell, his teammates, and Coach Monk Younger won some big ones. And Joseph H. McConnell . . . Phi Beta Kappa, president of the senior class, and later a U.Va. coach while putting himself through law school . . . has just kept on winning the big ones.

For this man, another big victory came 24 years after his exploits on the gridiron. In 1950, as chief of NBC, he appeared before the NCAA as spokesman for network television. Saturday's games had begun to air on the tube and the public was fascinated.

Ah, but the NCAA was considering barring TV from its games. Comparing the situation to the early days of radio when the colleges outlawed football broadcasts, McConnell pointed out that in 1949 attendance at the college games had risen four percent. And that at professional games — not televised — the gate was off nine percent.

McConnell, on behalf of the industry, offered to finance a study of the effects of TV on sports box-office receipts. And you see where we are today, able to view as many as four college contests on a given fall Saturday.

While at the summit of an impressive and long career, McConnell was a natural hero. The kind of hero that Bernard Malamud had in mind when told the fable known as "The Natural."

"Without heroes," wrote the author, "we're all plain people and don't know how far we can go."

McConnell showed us exactly how far we can go.

Spring 1991

McCormick Field

McCormick Field, oldest ball park in minor league baseball, has been replaced by a classic revival acclaimed by fans, government officials, and the sport's insiders alike. The incarnation of the original McCormick Field, home of the Class A Asheville (NC) Tourists of the South Atlantic League, has risen from the precise location of its origin in 1924.

"Up front, the local governing authorities told us they wanted two things above all," said John S. Ellis of Bowers, Ellis, and Watson, the Asheville-based architects of the new stadium. "They wanted to keep the park as an anchor to the historic downtown of this mecca of tourism, and they wanted to preserve the unique flavor of that great old park," Ellis added. "Home plate was to remain exactly where it had been."

There was another challenge, according to the architects: Bring the project in on time — that meant 210 days — and on budget, at $3,000,000 plus. At a capacity mark of 3500, that called for a per seat cost of $900.00.

The venerable layout is the proud possessor of "the most beautiful outfield in the world," according to the author of *Green Cathedrals,* a recent volume profiling all of the major league parks, those of the old Negro League, and some of the more exceptional minor league facilities. Philip J. Lowry, the writer, also observes that the vines on the stadium's walls "are even thicker than those at Wrigley," referring to the fan-friendly confines of the Chicago

Cubs playpen.

Set as it is among Carolina pines, oaks, and maples, Lowry says of this city's green cathedral, "It is a most beautiful park thanks to those lush surrounding trees and foliage."

Speaking of the ambiance at McCormick, architect W. Mike Watson comments on design features thusly, "As we saw the stadium, it was both a social and a sporting event site. Not only did one go to see a baseball game, but to see old friends and meet new ones."

"Current stadium design practices do not encourage this at all. One must go directly to their section and sit down. 'A walk around the ol' ball park to see who's here' would not be possible. As a result, we brought this concept of 'see and be seen' to the committee and the central concept for McCormick II was given new life."

(The concept sounds to one writer as though folks in this community sought to carry through on a vision of "the-city-as-village," a place where neighbor seeks out neighbor while on a modern-day gambol around the village green.)

"The old McCormick I crosswalk was woven back into the scheme of things," said Watson. He added that with this decision made, the remainder of the project took shape, with brick and concrete selected for longevity and ease of maintenance. Also space was set aside for future expansion in capacity and a standing seam metal roof chosen for durability and color.

A number of sportswriters and baseball administrators have remarked on McCormick II's ambiance as being in a league with that of the major league Baltimore Orioles Camden Yards. That one, too, is one of those "Best-of-the-old/Best-of-the-new" type green cathedrals fans and purists alike prefer.

In addition, as with virtually all publicly-owned sports palaces these days, McCormick II had to meet exacting governmental standards as a project. "With a timetable like we had," says Buncombe County Manager Steve Metcalf, "the architectural firm needed to provide a superb degree of client service. Around the clock. And that they did."

Jim Roach, Buncombe County's top man for finances, reminds, "The numbers were absolutely crucial. Because about the same time as the old place began to deteriorate rapidly, we had some other 'big ticket' items on the agenda — a new jail, another landfill, to mention just a couple. But the fact was we didn't want to lose professional baseball in these mountains."

How have fans responded at the gate? "Very well, indeed," General Manager Ron McKee declares. The home team finished third in attendance in a field of 14 league entries last season, topping clubs in larger cities such as Columbia, South Carolina, and Augusta, Georgia. The league as a whole drew over a million-and-a-half . . . more than several individual franchises in "the bigs."

"We should be up fairly substantially," McKee allows while commenting on the increasingly robust appetite for the national pastime around the nation, almost uniformly. A great deal of international interest is apparent, as well.

McKee himself is one with a feel for the numbers. His operation was selected by *Baseball America,* it is worth noting, the best managed in all Class A professional baseball for 1990-1991. He is aware too of the Tourists' impact on community.

And when it comes to consideration of factors governmental managers look at when minor league baseball's local impact is considered, this city with a metro population of almost 180,000 seems to fit the desired profile for success. For instance, a survey by the sports coalition of the International City Management Association reports:

- Local government officials in communities agreed that the presence of minor league ball in their areas is important. (85.1%)
- A majority of those reporting — like Asheville —

are without any major league sports, other minor league teams, or big-time college sports. (60.2%)

— The officials considered their teams "a significant entertainment opportunity." (98.9%)

Those polled cited factors being derived from the sport: Economic benefits, 84.1 percent; regional prestige, 83 percent; community identity, 83 percent; and civic pride, 78.4 percent.

In addition, the attraction of non-residents to their communities — a very big plus for Asheville, the tourism hub — was rated high by 72 percent. Some of that number see professional baseball at this level as a specific component of their economic or community development strategies. They see the clubs as generating "media exposure" and improving "quality of life."

Call it ambiance or label it economic impact. Even think of it as the perfect entertainment for a mid-sized city like this one. Local writer Bob Terrell sees it this way, "I've covered the Tourists right here since 1949, and at opening night for the new McCormick all I know is that an eerie feeling came over me — this was the very same place I had come to love so much. Only, if anything, the whole experience was even more enjoyable!"

Tourists boss McKee in agreeing with Terrell emphasizes his personally developed slogan summing up the McCormick Field experience: "It's just 'The Greatest Show on Dirt!' "

Folks in this skytop city are more than a little proud of the character of this gem of a ball park. So are the hometown heroes who made it possible. Mudville, this town is not. No one has struck out, and tonight at mighty McCormick, well, there is joy all about.

Horace "Bones" McKinney

Well, was there ever another one like Ol' Bonesy? Now I ask you, didn't the Lord break the mold when He made Horace Albert "Bones" McKinney?

Horace "Bones" McKinney is a one-of-a-kind treasure not only for Wake Forest fans, but for his many friends throughout the Tarheel state. The former Wake net coach is in great demand as an after dinner speaker.

When Bones resigned from his job as head basketball coach at Wake Forest in 1965, most folks thought the ACC was losing not only a fine coach, but the conference's most colorful character.

No way. At last weekend's ACC tournament in Charlotte, there he was, cuttin' up and carryin' on again, telling those great stories on himself.

The Charlotte Observer reports a couple of tall tales this way: In a game with Bill Bradley and Princeton, "Well, sir, I was wearing loafers and one of my shoes flew off and went almost to midcourt. I ran out there to get it and then my pen fell out of my shirt pocket. They were coming back down the court by then so I just stuck the shoe under my arm, threw up my hands and started playing defense."

A devoutly religious man and an ordained minister, Bones McKinney still makes talks all over this state, often to church youth groups. And he writes a newspaper column about life and its moments of humor and compassion.

Once he was hit with a technical foul and immediately had words with the official. The official said, "I thought you were a preacher!" Bones fired right back, "And I thought you were a referee!"

McKinney posted an eight-year record of 122 wins and 94 losses. He was a great player at State and Carolina. Then, too, he had a successful career playing in the NBA.

As a coach, however, McKinney took the Deacons to five straight conference championship games, claiming the top prize in '61 and '62 . . . Wake's only ACC titles. His '62 team won third place in the NCAA, scoring a season-ending win over UCLA.

Bones McKinney — past and present — brings something very special to the college game. Something that it fairly cries out for in a day of taking itself all too seriously.

What's that, you ask? A sense of humor and a sense of perspective.

This man, McKinney, usually wins the storytelling title at tournament time. But not this time. That award goes to one-time ACC official Lou Bello of Chapel Hill.

"Bones was really riding me," Bello said. "He was saying, 'How about a call for Wake Forest? Just one call for us. You can't even see what's going on, Lou. You don't even know what the score is!' "

The ref heard McKinney's last line and without missing a beat glanced up at the scoreboard and said, "Yes, I do know the score. We've got you by four!"

Winter 1991

John H. Moss

John Moss is comfortably settled into his box seat directly behind home plate on opening night at the glistening new ball park of the Class A SALLY League's Asheville tourists. His eyes sparkle as dozens of festive plumes erupt during the pre-game fireworks display.

A veteran of almost a half-century of leadership in the world of baseball's minor leagues, he has witnessed hundreds of such opening night extravaganzas. But the enthusiasm and the energy that flow from the man from King's Mountain continue to be rekindled each time an ump cries, "Play ball!"

This particular campaign in 1992, however, finds the-President-for-Life of the 14-team South Atlantic League of Professional Baseball Clubs poised at perhaps the pinnacle of a long and distinguished career. He serves as chairman of the executive committee of all organized baseball. He was named "Mr. Baseball" for 1990-1991. And he has brought his home state more franchises than any other in the union. More even than Texas or California.

So it's symbolic that on this night Moss presides over the dedication of the facility that replaced the oldest in minor league baseball. One that was the fourth-oldest in all of baseball, historic McCormick Field.

Flanking him here — in what former announcing great Mel Allen used to call "the Catbird's Seat" — are the general manager of the Houston Astros and his farm director. As well as the chairman of the local county commission.

By this juncture in the young season (Friday, April 17), the league president has overseen a half-dozen such openers. He has conferred with countless city delegations, all vying to become the SALLY's fifteenth and sixteenth entries. Five of fourteen franchises have erected new stadiums in just the past year.

Each of the cities with a new layout is rightfully proud. None more so than Asheville, with a park that still nestles comfortably into the same pines that — during an era before the fences — hid spare balls with which outfielders threw out astonished base runners trotting around the bases after hitting "home runs."

Architect Mike Watson of Bowers, Ellis, and Watson of the skytop city points out, "Just look at our cost per seat — that's what makes our local elected officials happy. And that's what has prompted other cities to enquire if the feat can be replicated there."

Attendance, moreover, promises to hit the two million mark in the near future for the SALLY League. So say some savvy students of the business side of the game.

"That's an indication of the great and enduring popularity of The National Pastime, especially in our region," says Moss, a former infielder. (He confesses to signing a pro contract before age 16, adding that he was a "good field-no hit" performer who soon realized his future lay with the administrative side of the game.)

Baseball always has represented this man's particular "Field of Dreams," to borrow the title of Hollywood's recent blockbuster movie. Remember the one about the Iowa farmer who constructs a diamond amid the cornfields for the spirits of old-time legends to come and play?

The character, played by Kevin Costner, passionately believed when the voices of departed heroes echoed about him, "Build it and we will come. . . ."

One speculates that John Henry Moss, too, hears such inner voices. That they whisper to him to build. And to keep building.

Whatever the case, the result hasn't been too bad for a good field-no hit infielder with a dream. A dream he shares on countless Southern nights at the ball parks with others — almost two million others.

So, to John Moss we would say: You have built this league, John, and we have come. Yes, we have come. . . .

Spring 1992

John Henry Moss (center), president-for-life of the Sally League, opened up the Carolinas for an unprecedented number of minor league baseball entries. His most unforgettable mentor: Branch Rickey, "The Mahatma," whose Brooklyn Dodgers broke the color line with Jackie Robinson. He is pictured here at opening night of the Asheville Tourists' new McCormick Field.

Rock Norman, Nath Thompson, and Their Happy Campers

Nath Thompson (second from left) and Rock Norman (third from left) of 68-year-old Camp Carolina for Boys represent the kind of leadership that has made the state home to 55 private summer camps for young men and women. (Others in the photo are Hoke Sloan, far left, and Dick Barnard, far right.)

For 68 summers now boys have been gathered up and delivered by parents to the rough-hewn office beside the lake at Camp Carolina, outside Brevard, North Carolina.

In that setting, they are introduced to 224 wooded acres nestled among Transylvania County's picturesque coves and cataracts.

Through the years they have come, adventurous (and sometimes mischievous) city boys, as well as small town boys eager to prove themselves in this sylvan environment.

"I call 'em 'city cool' and 'country cocky,' a lot of 'em," says Nath Thompson, current director of Camp Carolina. "There are some who are just plain shy," explains the personable Thompson, a one-time University of Georgia basketball player. "But we like to think they all have benefited from the self-reliance, teamwork, and love of nature they learned here."

With that as the camps' legacy, it's no wonder that new generations of campers are attracted annually to the 55 camps for boys and girls in the state. Neither is it surprising that the camps were founded by — and to this day are largely run by — teachers and administrators, college athletes and coaches. (When I once asked my late father, as one of those who loved camp work, why he remained in school work, his answer was persuasive: "Three reasons, son — June, July, and August." There was no summer school back then.)

As an example, now-defunct Camp Sapphire was run by the co-head football coaches at UNC, Bob and Bill Fetzer. And one summer in the 1920's, The Brothers Fetzer turned the conduct of sessions over to the famed "Four Horsemen" of Notre Dame. Camp Sequoyah brought in "Dad" Miller, a prep school coach of mine, along with others from the ranks of area educators.

Professor D. Meade Barnard of R.E. Lee High School in Jacksonville, Florida, was an early leader at Camp Carolina. His son, an official at Davidson College, would follow. Another pioneer, Dean J.M. McConnell, hailed from the same institution, as did football coach Monk Younger and gridiron center Joseph H. McConnell, son of the dean. (The son went on to assume the presidency of the NBC

Network, where his became the first network to televise NCAA football contests; he owned Camp Carolina until 1981.)

Other "founding fathers" at the Brevard camp were Hoke Sloan (tennis coach, Clemson), Captain North Pritchett (AD at Virginia), and Dr. Taliaferro Thompson (Union Theological Seminary). Tennessee supplied Frank Emmannuel, All-America linebacker, and track coach Chuck Rohe, now executive director of the Citrus Bowl.

Leonard Dick, athletic director at Woodberry Forest School, signed on, as did M.B. Chafin, tennis coach at Florida.

Arguably, however, the cornerstone of the venerable camp's great prominence — lifting it to the rank of other in-state institutions as Camps Mondamin and the now-dormant Sequoyah — was the late A.W. "Rock" Norman, coach of basketball and baseball at Clemson. What was the key to Norman's enormous influence, I asked Nath Thompson (whose wife, Mary, and elder son Charles now help him run the place).

A camp administrator for almost 40 years now, Thompson allowed as to how, "Campers loved Rock but, just as important, they respected the man." He emphasized, "There never was a better motivator of young people; years after you left camp you would realize that Rock had inspired you to put forth a great effort — and that in doing so you had grown up a lot."

As one might expect, the more established camps have revered traditions. (A number of Thompson's charges, for instance, have third and fourth generation credentials.) One of those at the Brevard layout is the awarding of "The Order of the Old Man of the Mountains." Its object is to recognize good citizenship in the cabins, as well as good attitude and participation in activities.

Fact is, there was a time when my then-eleven-year-old daughter was offered a trail ride by old college mate, Nath. The little city girl from Ohio, equestrienne that she is, leapt at the opportunity. After the ride, as we made our way back to Asheville in a downpour, she went quiet on me. Then Dallas asked pensively, "Dad, that Mr. Thompson, *HE'S* really 'The Old Man of the Mountains,' isn't he?"

Old, the man may have seemed to that pre-teen; and maybe wise, as well. The observation somehow prompted recall of some lines penned as the pioneers surged westward: "Give me men to match my mountains/ Men to match my plains/ Men with empires in their purpose/ And new eras in their brains."

Yes, men to match our Carolina mountains. Men like Nath Thompson, Rock Norman, and others of their ilk in camping. Their influence upon young boys has been considerable.

But, of course, as part of their rites of passage these young men come to know and admire the young women attending the many sister camps in the Old North State – among them, Rockbrook, Greystone, Keystone, and Merrie-Woode.

And it's safe to conclude, given the proper introductions, these happy campers will begin to contemplate an altogether different kind of "mountaintop experience." One hand-in-hand, say, with a life's companion.

Oh, boy! That's life for you.

Winter 1992

Gaylord Perry

Gaylord Perry, baseball Hall of Famer, won more than 300 games in the big leagues. Perry always "had something extra on the ball," according to some observers.

Gaylord Perry, the record will show, won over 300 games in the big leagues. He also was the first pitcher to win baseball's coveted Cy Young Award in both leagues.

The big right hander from Williamston, North Carolina, coached baseball at Limestone College down the road in Gaffney from 1986 until just recently. And on January 8, 1991, The Baseball Writers of America voted Gaylord into The Hall of Fame.

It was a fitting reward for a man who has done it all — everything from lining off the field and buying uniforms for the team at Limestone to spreading the word among young folks about the dangers of drugs. And this is the man who heard an area radio sports commentator in early January call for his induction into the hall after he came up just short of entry in the two previous years.

The announcer had said, "Perry always came through when

the chips were down," and deserved fans' support, "because he's one of us." Then came a call from a Limestone booster, John Childress, who asked for a tape of the commentary to be used in a video being produced to solicit support for a sports complex bearing the old pitcher's name at the college. Gaylord had been touched by the comments, and, sure enough, he was voted into the hallowed confines of Cooperstown three days later.

It was thirty-three years after the San Francisco Giants had signed him out of Williamstown (NC) High School. And what a glorious career it was!

A few high points: 1962 — Made the Giants major league roster; 1972 — Traded to the Indians, had a 24-16 season, won the Cy Young Award; 1974 — Tied American League record, winning 15 games in a row. And there's more: 1978 — Traded to the San Diego Padres, had a 21-6 season, became the first person to win the Cy Young in both leagues; 1982 — Became fifteenth pitcher to win 300 games (with the Seattle Mariners that year).

The Tarheel native in 1983 passed the great Walter Johnson to capture third place for all-time strikeouts.

Those are a few of the stats. But here's what I consider the real measure of this athlete. A few years ago the-big-time-pitcher-turned-small-college-coach initially accepted an offer from a leading brewer to appear in television spots. (You know, the old jocks screaming "Lighter," "Less filling.") Days later the coach called to say the deal was off. He couldn't go through all that "and then tell my players to lay off that stuff."

Yes, I was the announcer whose voice Gaylord Perry heard pulling for his induction into the Hall of Fame. And it was my turn to hear a soft and humble voice on the telephone a little later.

"Thank you, Bubba," the voice said.

Now you just know Gaylord Perry is a Tarheel original. He's one of us, and he's done us proud by the way he's played the game.

Winter 1992

The Reverend Doctor A. Lamar Potts, Jr.

Country singer Loretta Lynn is supposed to have said it on one of those afternoon television talk shows: "Merv, life is lived the way these songs is wrote."

One of the better college athletes of my day doubtlessly would agree on that score with the country music idol though he would likely have phrased the sentiment differently (and a tad more grammatically), as he is now The Reverend Doctor A. Lamar Potts, Jr. of Columbia, South Carolina.

To me, though, he is still good ol' "Bubba" Potts, Georgia basketball forward and ace of the baseball pitching staff. Even if he is the senior minister at a very large Presbyterian church in The Palmetto State's capital city. And one, I am compelled to add, whose doctoral dissertation dealt with theological themes implicit in the country music of his homeland.

Yes, Reverend Potts has his roots quite deep in the neighboring state's red clay, where his father was a county sheriff and Bubba was the four letterman down at the local high school. And downright bashful though he was, Potts also patiently wooed and ultimately wed one of the prettiest girls ever to come down the pike back home. Miss Ann Manget, she was.

And if Ann Manget wasn't the prettiest girl in that little town, her twin sister, Jane, was. (Just to think, too, Jane was my sweetheart once — however briefly, I must add by way of caveat, since no one should suffer such a stigma on a long-term basis.)

So, the sports hero and the hometown beauty, huh, another American success story? Not exactly. Not the conventional kind, at least.

Let's start in college. Already ranking high as a draft prospect among baseball's major league scouts, the one-time hurler literally threw his arm out. Someone dared the young pitcher one spring evening to throw a tennis ball over the Coweta County Courthouse. ("Ever see a cow eat a county?" That's what Sister Jane would ask me teasingly.)

End of a promising career in professional baseball.

After the future clergyman starred on the college basketball team, Bubba and Ann tied the knot. As I recall, a hitch in the Air Force followed, then it was back to Newnan, Georgia. Potts formed a highly successful ready-mix concrete and construction business. But the bottom dropped out.

Just as surely as it always did for Bubba and his Bulldogs when they took on Baron Rupp's Kentucky Wildcats. (The Baron was one to sniff, "Playing the University of Georgia in basketball is about as exciting as kissing your baby sister." And sniff indeed is the right word, for Rupp always wore what I would call a smell on his face.)

Well, deep in debt and approaching age 40, Lamar Potts, Jr. and his bride made a joint decision: Back to school — to seminary and preparation for the ministry.

But this time success came in a different package. In the form of satisfaction, service, and — well — a kind of inner serenity. And with it all, like many of those whose waters run still and run deep, the Reverend Potts was one who always had listened carefully for the deeper meanings in the words and songs of his neighbors.

In one country lyric, says Potts, the singer laments a loveless marriage with, "When we said I do, we did, but you don't anymore." In that he hears a call for compassion and with it the biblical charge, "Weep with those who weep." We commiserate as well with one whom life has seemingly passed by when he sings, "The future's not what it used to be."

The clergyman cites the verse, "We sat down and wept when we remembered Zion," as epitomizing the Southerner's penchant for "remembering and longing for home." Randy Travis' "Storms of Life" sounds that theme clearly, he says, as a wanderer sings: "I left my soul out in the rain/ What a price I've had to pay/ The storms of life are washing me away," — and then pines for home and a long-ago-abandoned, more committed way of life.

When Bubba Potts makes a connection in his preaching between the singer's pain and the listener's compassion, we feel moved to join in the chorus with Hank Williams, Jr. "How I love that hurtin' music," Williams wails, " 'cause, Lord, I'm hurtin' too."

How has this contemplation of country music and the lot of the common man guided this man of the cloth in the pursuit of his calling?

Well, as I recall it, Lamar Potts, Jr. never was an athlete blessed with blinding speed as a runner. (In one high school playoff game, I saw him haul in pass after pass, only to rumble three or four more yards at best after the catch. But of course it took several piggyback riders ultimately to topple him.) I guess that recollection is why I asked him about the biblical injunction, "Let us run with patience the race set before us, looking always to Jesus, the author and finisher of our faith. . . ."

I could have predicted his answer, "Yep, country music has something to say about that, too. You know, we're told that our job is to stay the course, not play to the crowd."

With the patience of the long-distance runner — and a practiced ear for hearing the noblest impulses of God's ordinary folks — this shepherd of his flock is now moving rapidly ahead on the backstretch. And after our first visit in over 30 years, I think I at last can see how the course he chose to run might have been infinitely more rewarding. A lot more meaningful, say, than a handful of big league seasons in the sun.

In the long run, that is. The really long run.

Winter 1992

Lamar "Bubba" Potts was a superior athlete and, for a time, a successful businessman until a call came to the ministry. A promising pitcher who threw his arm away, Potts wrote his doctoral dissertation on religious themes implicit in the country and western music lyrics of his homeland.

Jerry Richardson

He's a native of Spring Hope, North Carolina; won Little All-America honors in football at Wofford College in Spartanburg, South Carolina; and was a target for Johnny Unitas' aerials with the old Baltimore Colts. Now he's on the brink of bringing an NFL franchise to the Carolinas.

Jerry Richardson, turned down for a $250 salary increase by the NFL's Baltimore Colts, then became "a fry cook" at the very first Hardee's franchise. Now the founder of the nation's fourth-largest food service empire, Richardson is intent on bringing an NFL franchise to the Carolinas.

That's Jerry Richardson, president and CEO of the fourth-largest food service outfit in the country, TW Services of Spartanburg. And when the Colts refused to give him a $250 raise, in 1961 Richardson started flipping burgers for a living.

If the man instead

had turned to coaching, he would have been a kind of hands-on tutor — one given to demonstrating physically for his charges the finer points of the game. At least that's my impression from reading a recent *Sports Illustrated* piece on the fellow.

Charlotte megabanker Hugh McColl, who loaned the one-time receiver the money to get his business launched, apparently agrees. Says McColl, "Here was an operating officer who could cook, serve customers, and clean up the trash. He led by example."

Richardson's Fayetteville High School coach, Bob Prevatte (now of Gaffney), echoes the leadership theme. "He was that way as a player," says Prevatte, who later would join Richardson in the Hardee's business.

"In the business world, when Jerry's employees have to work holidays, he typically shows up to cook or to clean up the parking lot. That's leading by example alright!" says the old coach.

Still does, three decades after he and partner Charlie Bradshaw — Richardson's teammate and battery mate at Wofford — obtained the very first Hardee's franchise. Now his company owns 498 of them — plus a flock of Quincy's Steakhouses and the entire Denny's restaurant chain. Too, the Spartanburg concern operates a passel of quick service chicken places named El Pollo Loco. (I think that means "The Crazy Chicken.")

Jerry Richardson's combine also is quite big in the contract food management field, vending and the like. Among his Canteen unit's accounts are four major league baseball parks and a like number of NFL layouts. Not to mention several national parks, including Yellowstone.

So the 55-year-old Ex-Colt has been around ball parks for a while now. And it's natural, given the instincts of an overachieving 185-pound competitor, that he'd want to put his managerial skills to the test by landing an NFL franchise. He points out, incidentally, that within a 150-mile radius of his proposed stadium in Charlotte there are 9.6 million people.

Lots of numbers here? Uh-huh, the man is savvy with the numbers, as well as high finance and big-time risk taking. After all, his TW Services did $3.7 billion (repeat, billions) in business last year. Not bad for a founder who started out with only his share from the Colt's 1959 championship purse. That was $4,674.

Ah, those '59 Colts: People like Unitas, Raymond Berry, Alan "The Horse" Ameche. Jerry Richardson and Art De-Carlo. Art De Carlo? Yep, my college classmate was the equivalent of a baseball utility man. He could play four positions in that or any other season, which helped him survive the final cuts for about six seasons. And, lucky me, I spent that '59 campaign right there in Baltimore while in a training assignment with a giant soft drink company.

The route salesmen at work were suitably impressed by my connection with the popular De Carlo. Especially Melvin, whom I suspected of being a part-time bookmaker and who always wanted me to ask De Carlo questions about injuries on the team. ("Can't do it, Melvin, or 'Big Daddy' Lipscomb says he'll transform him into a permanent part of the masonry at Memorial Stadium!")

Oh, I took Art De Carlo and his new bride out to dinner at a great old German place, Haussner's, and picked up the tab. After the meal, I grandly ordered cherries jubilee. You know, with the flames and all. Still, oh boy, no tickets to the championship game. (Melvin offered to make me a permanent part of the plant's parking lot on Hamburg Street.)

After that epic struggle with the Giants in 1959, I got a call from De Carlo. "Get this," he says, "Our buddy Bratkowski (Zeke, the Bear's quarterback) called to ask how we liked the automobiles Rosenbloom gave us. Hah, he doesn't know Carroll Rosenbloom!"

Well, the Colts owner was known for being kind to his players, but was hardly the sort for that kind of generosity. Fact is, the club wouldn't even give Jerry Richardson a $250 raise for the next season. So the backup receiver became an enterprising fry cook.

Because of his efforts, we may have our own NFL team around here. So thank you, Mr. Richardson, and thank you, Mr. Carroll Rosenbloom, for setting it all in motion 30 years ago.

As the Russians now seem to realize, American free enterprise can bring a whole lot of good things. If, that is, you have the right brand of leadership, are willing to work hard, and have the gumption to take risks. Like "Old Number 87," J.J. Richardson.

Winter 1991

(L) Hardee's 15¢ hamburgers, 10¢ French fries, and 10¢ and 15¢ soft drinks were a big hit instantly. Pictured above are teenagers in Shelby, NC, during 1963 enjoying soft drinks and Hardee's hamburgers at one of the original walk-up Spartan Food Systems restaurants. Today, the company operates three Hardee's in Shelby. (R) Jerry Richardson.

Sanders Rowland

Sports marketing they call it, and it's the hottest thing going in competitive venues all across the nation. One company vies for the designation, "official beef stew of the 1992 Olympics." Another capitalizes on its connection with a major motor sports event, placing life-sized cutouts of driver Kyle Petty in convenience stores.

There's the Kemper Open, not to mention "the Mazda Gator Bowl." But there was a fellow named Sanders Rowland who could have taught all these marketers a few things, though. The Asheville resident, you see, was the first man to market soft drinks at a college football game.

Rowland, who died last October at age 83, didn't stop there. When football's Paul (Bear) Bryant came back to alma mater Alabama almost three decades after Rowland's precedent-setting gridiron feat, the man was there to sign him to a promotional contract.

"Great Pair Says the Bear" billboards proclaimed all over Sanders' territory as they sold "Atlanta Holy Water" by means of a tie-in with a regional potato chip outfit. Later on, the same executive also encouraged his Western North Carolina bottling entry in establishing the annual Southern Conference Basketball Tournament in Asheville as a premier sponsored extravaganza featured on the ESPN Network.

In a gesture having little to do with pushing product, the same bottling company's jet would fly the late Western Carolina football coach Bobby Waters to Houston for regular therapy sessions. Waters was suffering from Lou Gehrig's Disease.

Every high school sports facility the soft drink purveyor could reach had Rowland's red scoreboard — and exclusivity for his drinks. It was a far cry from the Saturday in fall 1929 when 60,000 fans at the Harvard-Army game drank only fourteen bottles!

Admittedly, some of the schemes hatched by *Papa Coke* (the title of Sanders Rowland's book with Bob Terrell) just didn't cut it. One idea was pouring the stuff over cornflakes at breakfast, as he himself was wont to do.

During a 66-year career in his industry he had stops in New Haven, Boston, and Hartford. Add to those Providence, Pittsburgh, and then Birmingham. Fidgeting after retirement, he moved on to Asheville to jump start that operation out of the red.

Always the showman, the portly promoter often originated radio broadcasts of The Big Bands live from his headquarters city to spur on the war effort. And when gasoline was in short supply in World War II, he had horse-drawn wagons deliver the product on the home front.

Last fall, the strain of a sportsman's life-style, combined with cola wars pressures, finally took their toll. Rowland was being attended by a nice lady in a convalescent home. She wrote that Papa was looking forward to my visit when I returned from Seattle to my Carolina home in the fall.

I came a few days too late and even missed his funeral on October 26 at the church I had attended with the old pro. Just a few short months before, we had enjoyed a reunion at "Sneakers," a popular restaurant on Hendersonville Road near Mimi and Sanders' place. I was home for the holidays.

Papa had asked, "Gull, Gull?" His speech, slurred by strokes, made it impossible to understand the question. Tears of frustration came to his eyes, and I later realized that he was asking about my 14-year-old daughter in Ohio who had visited at Sanders' home once and came under the old charmer's spell. ("I can't believe THE AUTHOR gave me an autographed copy of his book!")

The top brass of the Atlanta giant descended on "The Land of the Sky" to pay their respects, as did countless other friends and admirers. As they hoisted their glasses later that evening in tribute to the "Great Salesman," many wondered out loud who would replace him in competing for refreshment business at sports events. Papa had laid down his marketing trophies and gone home.

Well, Sanders Rowland probably has another exclusive up there by now. Some salesmen radiate such charm and goodwill that, like him, they simply can't be turned down.

Sanders Rowland, "Papa Coke" to his legion of friends in the soft drink industry, was the first to sell the drinks at a college football game. He was a premier sports marketer, according to friends, recording many other "firsts."

Sam Schuman

Picture the scene in the Broadway hit, *Foxfire,* starring Hume Cronyn and Jessica Tandy as the elderly mountain couple, Hector and Annie Nations. Theirs is "the Code of the Hills," their values much like those of many of our neighbors in these parts.

The aging Hector is annoyed in the scene by a young student journalist's questions about the couple's mores, their folkways. "You want me to say the old ways was better, don't you?" he carps.

"Well, some of 'em was!" Hector snaps. Later, on the subject of television, he snorts, "Biggest piece a' sin furniture they is!"

In the minds of some reform-minded college and university leaders — including UNC's Bill Friday — there is an inclination to echo some of Hector's views when it comes to the subject of athletics on campus today. Television, it is felt by many academic figures, has brought an unreal dimension to the college game. It has upped the ante to such an extent that only 50 or 100 superpowers in college athletics can compete at such a rarified level.

Too, at an institution like UNC-Asheville — with just 3200 undergraduates, but still pegged at the I-A competitive level of play — the school's new chancellor, Sam Schuman, is advocating something like "the old ways." You know, back when the college game was for true amateurs. A time when you came out for the team, you were issued a uniform. And when coaches actually taught classes.

The strong-minded Schuman's school actually ended up with a basketball road win over big-time Cal-Berkeley a few years ago. (Reminds me of State's Jim Valvano and his dictum, "Don't ever schedule those hyphenated schools early in the season!" Right, the results are occasionally embarrassing.)

But the larger question here is how UNC-A and its ilk stay at that I-A level when the USC's and the Notre Dames of the NCAA start insisting on more grants-in-aid, bigger stadiums, and a greater number of varsity athletic teams. Well, on that Schuman says it's time for some common sense and an affirmation of higher education's central mission.

The persuasive educator first shed light on the subject at a news conference several weeks ago after a university commission pondered its future directions on the playing field. Then this writer spoke again to Schuman after his customary mid-day run. (Yep, the man has competed twice in the famed Boston Marathon; "A sound mind in a sound body" — that kind of thinking — is a deeply held tenet for the scholar in English Literature.)

The stand UNC-Asheville has taken, in fact, is worthy, I submit of an allusion borrowed straight from that foremost student of the language, Sir Winston Churchill. The date was June 18, 1940, and England was at war.

The Prime Minister addressed the House of Commons: "Let us therefore brace ourselves to our duties and so bear ourselves that men [years later] will say, 'This was their finest hour.' " In his own way the chancellor at Asheville took his own courageous position on a burning question for our educational institutions and their athletes.

Here's the way Sam Schuman sees it:

> "I am mandating that UNCA not increase the proportion of our operating funds designated for intercollegiate athletics. That proportion is now between 2.5% and 3% — about one dollar out of every 40 we spend on anything. We cannot and will not fund sports programs by starving our library, laboratories and classrooms. I am confident our athletic

boosters will work enthusiastically as part of our UNCA team to help seek the support we need to maintain and heighten our excellence in the classroom and on the playing field alike."

He added:

"In all these matters — academic integrity and fiscal responsibility — the Coaching and Athletics Department staff are in complete accord with me and the Task Force. On the primacy of academics at UNCA, we speak with one voice."

Call it the Gramm-Rudman Act of college athletics. As with the Federal deficit, the matter of expenditures of funds needs a perspective. Just as the deficit needs a cap. At schools like Schuman's the cap makes great good sense — and the setting of it may represent "their finest hour."

Now, if only the networks who fatten the coffers of the really big boys — along with the athletic-shoe-contracts-for-the-coaches people, not to mention the professional agents — will kindly step back. . . . Well, maybe the college game can be played by amateurs interested primarily in a good education.

Now, about the standings, while in the hoop game UNC-A was ranked nationally about 294th among 300 rated schools by *USA Today*. Sam Schuman may take pride in yet another measure. At his place, students rank second and third, respectively, among those enrolled at the 16 State University System institutions. And a little while back, it was designated the only liberal arts university in the UNC System.

Even so, sports enthusiast that he is, Sam Schuman — the long-distance runner, remember? — admits he relishes the thought of an occasional victory over a really-big-time place like Cal-Berkeley. Or maybe another even closer to home.

Spring 1992

Frank Selvy

Frank Selvy of Furman is remembered for the night he pumped in 100 points against Newberry College, still an NCAA record. Selvy's high-scoring, wide-open style of play exposed many Southerners to the kind of game hitherto played only in Kentucky and Indiana.

In the 1940's and 1950's, basketball in the South was pretty much the stepchild of most high school and college sports programs, remember?

The game football's Bear Bryant scornfully called "bounceball" was something you did in the winter to keep in shape for spring football, or maybe track or baseball.

In my college days, hardly anyone went to the home games at leaky old Woodruff Hall, and Kentucky's incomparable "Baron" Rupp said, "Playing the University of Georgia in basketball is like kissing your baby sister."

Oh, that began to change some after so many thrashings at the hands of Rupp's Wildcats. Likewise, when Ev Case administered so many beatings to his Southern Conference brethren, Carolina included. At one point the NC State mentor had a string of 15 straight against the 'Heels.

Then in the early '50's, an old boy from Corbin, Kentucky, burst on the scene at Furman University, down the road in Greenville. What's the big deal about that? Well, consider this:

- On February 13, 1954, that Kentuckian scored 100 points against Newberry. Only one other Division I performer has come close, scoring just over 80.
- In that 1954 season – his last for the Paladins – Frank Selvy averaged 41.7 points per game, playing every minute of every game.
- And just so you don't get the idea that Selvy & Co. spent all their time whippin' up on the Newberrys and the Erskines. They whomped Georgia Tech 114-67 and manhandled Manhattan by 92-80 at Madison Square Garden.

Selvy ended up a three-year career at Furman with over 2500 points. And he captained the AP All-American Team in 1954. Moving on to a fine NBA career, "Fabulous Frank" at graduation time held 24 major college records.

After I saw this man Selvy perform, I never again thought of the game a mere off-season conditioner for football players. Frank Selvy changed that, and he made believers of us all, didn't he?

Spring 1991

Tommy Sheriff

It was a typical early season mismatch, the Division IAA Catamounts of Western Carolina taking on Division IA's University of Georgia Bulldogs. In the nation's sixth-largest college stadium, capacity 85,434.

The 'Cats opened at NC State last season, and the outcome was bad enough — 67-0 — but what was most embarrassing was that they failed to even record a single first down. In other initial outings during the last decade, the people from Cullowhee have been downed by Clemson, 44-10 (1983); by the Tigers again, 43-3 (1987); Boston College, 44-24 (1984); and State, 45-6 (1988).

Why would a school and its athletic department put their teams at such disadvantage by scheduling the boys from the big time? The answer came from the Catamounts' Sports Information Director, Steve White. "We got a $200,000 guarantee from Georgia, and that will go a long way toward financing our total athletic program," said White.

What could have been going through head coach Steve Hodgin's mind this hot, muggy Saturday as he sent forth his charges? Perhaps he borrowed a line from former Georgia Southern coach Erk Russell. Russell exhorted his men about to face nationally-ranked Florida State, "All right, go out there, kick off, and fall back to protect the equipment!"

Time was when little Presbyterian College — home to all of 500 students — traditionally took on powerful Clemson at "Death Valley." (As a matter of fact, it was PC coach Lonnie McMillan who coined that phrase, "Death Valley," in a pre-game interview.)

The year I attended the school, we managed to top Furman in old Sirrine Stadium in Greenville, 13-12. Then on to Clemson for the pleasure of a 77-0 humiliation, and a maiming demoralization of the Blue Hose program for the remainder of the year.

Ah, but finance the school's program from the guarantee, we did.

A friend who lived across the hall in old Spencer Hall — a 175-pound starting sophomore tackle, Tommy Sheriff of Orangeburg, South Carolina — well, he survived that blowout somehow. Then came the gut-wrenching news three or four years later: Tommy was killed just after the armistice ending the Korean Conflict was signed. A grenade struck him square in the heart.

There were some heroes who took the field for Western against Georgia, just as well. One was the starting offensive right tackle, Todd Pearson of Shelby. And Jim Taylor out at the high school, he can be proud of this old Pearson boy. He showed up, and as Woody Allen says, "Ninety percent of life is just showing up."

Well, that's Pearson. He's a lot like the fellow whose epitaph Harry Truman said he wished could be his own. The fellow's gravestone read: "Here lies Jack Jones; he done his dangdest." That's what Todd Pearson "done," his dangdest. So did his teammates.

Now, what with peace breaking out all over, maybe these Catamounts won't meet the same fate as my old college buddy, Sheriff. He was one of the bravest and most self-sacrificing athletes I ever knew, God bless him.

As for Western, they have a week off after Georgia. And wouldn't it be something if they went over to Winston-Salem the next week and knocked off the Demon Deacons of Wake Forest? They have a way of coming back from adversity, do those 'Cats from Western.

Sheriff came back a different way — in a wooden box. But teammates E.M. Watt, Joe Kirven, and the others on that Presbyterian team 41 years ago . . . they'll think about the old boy as football season gets started this fall. I know I will.

We'll realize that what he taught us is a lot of what life is about — and how fragile it is. Like Western, Tommy Sheriff did his "dangdest," and I'm proud I knew him.

Summer 1991

Tommy Sheriff of Presbyterian College typified the kind of undersized but gallant linemen who brought honor to their alma maters in the 1940's and 1950's. Lieutenant Thomas Myron Sheriff, still in his early twenties, was killed in a tragic accident shortly after the armistice ending the Korean conflict.

Canie Smith

Canie Smith (R) was dubbed "Pioneer" by a teammate in his high school playing days in Asheville. Smith later became a different kind of pioneer as one of the first American airmen permitted to fly behind what was known until recently as the Iron Curtain.

The upperclassman, a star on the high school's football team, eyed the young sophomore trying out for the varsity. The younger boy always wore jeans and those lace-up boots some of us favored in the 1950's.

"You look like some kind of dang pioneer," said the star to a 14-year-old Canie Smith. The nickname stuck. And he would become a 170-pound fullback flashing occasional brilliance for the Lee Edwards Maroon Devil elevens in those halcyon days. Later he lettered in tennis at Chapel Hill.

A pioneer he was, and once again this summer he proved himself another kind of pioneer.

He and UNC pal Dr. Ted Rogers flew their Cessna 210 to Nome, Alaska, and then on to a little city in the Soviet Union named Provideniya. The 7,000-mile flight marked the first invitation to a group of American pilots to fly into the USSR's previously-sealed-off confines.

The pair were unlike young Mathias Rust — remember the name? — who in 1988 flew across the Russian border undetected in his light aircraft smack into Red Square. No, Smith and Rogers had an official invitation to the USSR.

Rogers, a Waynesville opthalmologist, was first to learn of an opportunity to join the Alaskan Airmen's Association Russian-escorted flight to Siberia. The area men became part of a 35-plane, 80-person cadre led across the Bering Straits by a bulky old Aeroflot aircraft.

"As my wife, Lynn, had told me, 'Do it; it's the chance of a lifetime!' " said Smith, who heads a large wholesale drug company with operations in Raleigh and Asheville.

The pair were to stay in Provideniya in the home of a Russian host for four days and then return home. Soviet airmen were returning the hospitality shown them on a memorable visit to Seattle in 1978 or 1979.

In the little city — "About the size of Brevard," says Canie Smith — the North Carolina men were on the receiving end. And the Russians' attitudes "are equal to or better than our Southern hospitality — very warm and gracious," said the former three-term mayor of Biltmore Forest.

The birdmen from these parts were gone for two weeks and a day. And there were some anxious moments for them, to be sure. But the euphoria of landing on Russian soil made the voyage more than worthwhile. On landing, the friends popped out of the unpressurized cockpit, shaking hands and proclaiming, "Welcome to the Soviet Union, partner!"

All along the way, the twosome served as goodwill ambassadors for our region.

Oh, the old jock who pinned the nickname "Pioneer" on old Canie. Well, he's a one-time FAA controller in Atlanta named Jack Bishop. When he gets word of Smith's intrepid flight, there'll probably be an "attaboy" on the way. And about now, I'm wondering if maybe the two oughtn't rename the craft "The Pioneers."

Because that's exactly what they are. Just as much so as the Dan'l Boones, the Davy Crocketts, and the Bishop Asburys who staked out this land we know to be God's Country.

Summer 1991

Dean Smith

Dean Smith, son of high school educators, himself is a born teacher. Entering his fourth decade as cage coach at Chapel Hill, Smith's most enduring legacy at the school may be not "The Dean Dome," but a 98 percent graduation rate for his charges at the university.

How proud this state can be this weekend! Two. Mind you, two of our schools playing in the NCAA "Final Four."

The Atlantic Coast Conference sent six teams to The Big Show this year. And now both Carolina and Duke are there at the top again for the umpteenth time.

These are two of America's great educational institutions, and nobody, but nobody, doubts that. Great institutions, great traditions, great student-athletes. And two of the finest coaches in the business. But because Dean Smith has reached several more milestones during the course of this '90-'91 season, let's pay him one more tribute.

We all know Dean's stats by now: 30 seasons at the helm, only one losing season, 16 ACC championships, a national championship. And over 700 wins. The impressive list stretches on from there.

But listen up now. I submit that this one is the stat that should impress us most: A sparkling 97.8% graduation rate among Smith's Tarheel players. The Devils of Duke are up there, as well, at 90%. The national average stands at 56%.

Consider, too, what a rival coach once said about the Carolina program: "They're not a team of basketball players. They're a team of true student-athletes. That's why they've been so successful. They have quality people, not just quality athletes."

You know, this transplanted Kansan is the son of public schoolteachers and, to my way of thinking, Dean Smith is himself a teacher. (I'm one of those, as well, the son of teachers.)

Only lately have I begun to cherish that heritage — being the offspring of schoolteachers. I wish now that I had been one myself; a teacher, that is.

And maybe that's why, somehow, I think Coach Smith will appreciate what I'm saying here. He is now, and has been, a truly great teacher. And, by my lights, Coach Dean Smith of Carolina deserves an honorary degree in pedagogy — that's simply the highfalutin' word for teaching — yep, a doctorate degree from every institution of higher learning in the state of North Carolina.

And, like those Tarheel fans, I mean that, win or tie this weekend.

Well, that's our lesson for today. Class dismissed.

Winter 1991

Wallace Wade

Trying a little choice sports trivia on friends last week, I mentioned that lineman Fred Crawford of Duke was the state of North Carolina's first All-American in 1933. And I followed quickly with the observation that it was about time the Old North State had a Heisman Trophy winner.

(That train of thought invariably reminds us that Charlie "Choo Choo" Justice of Carolina – and Cherryville, North Carolina – should have won it, maybe twice, in the late 1940's.)

Well, it could have been that Crawford might never have earned his honor had it not been for William Wallace Wade's coming to Duke in 1931. What a prize: To land the coach from Alabama whose teams from 1923 through 1930 won 61 games, lost only 13, and tied 16!

Wallace Wade took the Crimson Tide to the Rose Bowl three times in eight campaigns. His 1926 squad was the first Southern aggregation invited to Pasadena, and 'Bama edged the University of Washington, 20-19.

He came close to working a similar miracle at Durham. But, boy, did he ever have a frosty personality!

As Edwin Pope, author of *Football's Greatest Coaches,* put it: "He stressed efficiency rather than theatrics, production rather than excuses. He saw, through ceaseless improvement of the single-wingback formation and the kicking game, that the job got done."

Yes, Wade was crusty and aloof. He was relentlessly resistant to change. And was he tough!

Pope tells us that at a halftime of a Duke-Carolina game, the coach asked end Earl Wentz what was wrong with him. "I'm sick," the player whined.

"Sick! Boy, get sick on Christmas Day. . . . Get sick on the Fourth of July. . . . Get sick anytime. . . . But don't get sick in the middle of the biggest game on our schedule!"

One day at practice, tackle Tom Rogers failed to rise after a pile-up. "What's the matter, Rogers?" The muffled reply, "I think I broke this leg, Coach."

Wade shot back, "Well, you've got another one, haven't you?"

But Wallace Wade could spur a player on to greater things, too. Before the Navy game in '48, the coach told tackle Al De Rogatis, "Today, you make All-America." A surprised De Rogatis recovered his composure in time – and make the dream team he did!

If the man's players were troops in warfare, then Wade was their supreme commander, General George Patton. In 14 seasons at Durham, Wallace Wade won six Southern Conference titles. His 1938 crew went undefeated, untied, and unscored upon. He had two Rose Bowl teams.

Then, according to the *Durham Herald's* Jack Horner, Colonel Wade returned from World War II a changed man. In his final few seasons, the spark was gone.

"I'm just not tough enough," said the old coach, shaken by the horrors of war. "The truth," wrote Horner – when Wade quit insisting on large, powerful recruits – "was that he had mellowed after seeing 135-pounders die for their country as gloriously as 200-pounders."

So, if as Sherman said, "War is hell," then Wallace Wade's Duke elevens must have been, well, the very devil on those glorious old gridirons of yesteryear.

Summer 1991

Wallace Wade took his Duke elevens to conference championships, Rose Bowl appearances, and seasons when his rugged defense scarcely permitted a score. Players acknowledged he was the supreme authority figure, not to be questioned on or off the field of play.

Clyde "Peahead" Walker

During the heyday of old-time radio — in the 1930's and 1940's — I remember a program called "Frank and Earnest." Get it, Frank and E-A-R-N-E-S-T.

Seems to me this pair were in the business of dispensing advice on the human condition, the common and ordinary problems of daily life. Does anyone else remember that show?

Well, today the program we spotlight is the "Frank and Peahead Show" — and the humor it featured was as broad as it was cornpone. The lead characters were, respectively, the head football coach at Clemson University and the head mentor at Wake Forest University, who later became the top man with the Canadian League's Montreal Alouettes.

They were Frank Howard of the Tigers and Douglas Clyde (Peahead) Walker of the Demon Deacons. Their "show" was augmented on occasion with appearances from Wallace Butts of Georgia and Herman Hickman of Yale, by way of the University of Tennessee.

The foursome weighed in at about a half ton. And these rather portly gentlemen liked and admired one another tremendously — though you certainly couldn't tell it by the repartee, the putdowns they engaged in.

They'd go away for so-called hunting trips. And the competition was keen to tell the tallest tale, eat the most barbecue, and — how can I say this? — drink the most firewater. It was said, too, that rather than lose a hotly-pursued prospect to an in-state rival, the potential loser would turn the prospect's dossier over to one of the other three and, figuratively, say "Sic 'im!"

I can hear it now: "Oh, no, you don't, Peahead, I swapped you Skudlowski last year and now you flat OWE me that linebacker from Moncks Corner!"

Before Wake moved to the attractive new campus in Winston-Salem, Walker would fly in talent from prospect-rich Pennsylvania to show them "our campus!" Driving them from the airport through the environs of Chapel Hill, North Carolina, Peahead would see off the high school hot-shots who left with glowing impressions of Southern charm, gorgeous coeds, and springtime in Dixie.

In 1959, Howard invited his pal Walker to appear on his television show prior to the Sugar Bowl game with superstar Billy Cannon and LSU's famed "Chinese Bandits." Peahead was uncharacteristically quiet. But before the show he had taken a moment to privately advise the show's host to turn to him, Walker, for the final comment.

The announcer obliged and asked, "Coach Walker, who's your pick to win this tremendous contest?" He drawled, "Wal', Clemson has a heap of talent, but I gotta go with the Chinese Bandits to win 'cause Clemson's got a Mongolian idiot for a coach!" Switch to commercial, fade as a sputtering Frank Howard is seen rising to throttle his erstwhile friend.

Of course, there was the time that coaching legend Bear Bryant of Alabama was diagramming a defensive set at a coaching clinic and inadvertently drew 12 "x's" on the board. A twangy voice rose from the back of the room, "Now I see why you never would give me a game, you blankety-blank, you'd have to play 12 MEN on defense to beat me!"

Before composing this piece, I didn't even check out Peahead's won-loss record. It was good, that much I knew.

I just remember that he left us with a smile on our face. And that's the way it was back then, fun. That and characters who made friends enough to last a lifetime. But as the late, great UNC publicist, Jake Wade, used to ask, "Say, what'd you want in a head football coach, Sir Galahad?"

Winter 1992

Bob Waters

There are few heroes who seem to represent all the qualities we admire. But one such was the New York Yankees' Lou Gehrig.

Remember the movie, "Pride of the Yankees," with Gary Cooper cast as the handsome first baseman? The slugger was the victim of a crippler we now call "Lou Gehrig's disease."

You know, we have our own special hero in Western North Carolina. One who died of that same dread malady. His name was Bob Waters, and he was the most beloved and successful football coach in the history of Western Carolina University. Unlike the Iron Horse of the Yanks, however, he was not allowed to exit with the same dignity as did Gehrig. That

Bob Waters led his Western Carolina Catamounts to gridiron prominence in NCAA Division IAA competition. Lou Gehrig's disease would fell the Cats' greatest leader.

after 20 golden seasons at the school.

The WCU chancellor — acutely aware of Waters' condition — asked that he step down. Then he replaced the pain-wracked coach, who wanted just one more season at Cullowhee.

How was that done? In a memorable column, Larry Pope, former sports editor and now executive editor of the *Asheville Citizen-Times,* wrote that a messenger was dispatched to the practice field, where he stuck a note in Waters' pocket.

A humiliated Bobby Waters had a response to the note-in-the-pocket incident. A few months later, he just quietly departed this world. The world of sports was stunned. There was an outpouring of grief and then outrage at the way this proud man was treated.

In 1960, I was in Los Angeles for a convention. In a contest at the Coliseum, the Catamounts' great leader — fresh out of Presbyterian College in the upstate — came in to replace the Forty Niners' Brodie and Tittle in the shotgun offense. He put on a veritable aerial circus to upend the Rams.

It was 27 years until I saw him again, and it was a tragic sight that greeted me. There he sat on the sidelines on a platform in a wheelchair. He didn't even wear headphones to speak with assistants up in the pressbox area. His limbs hung lifelessly from Waters' slender trunk.

By then, Waters was being flown almost weekly in a Coca-Cola jet for treatment at Houston's Baylor Hospital.

Fred Brown, dean of the region's sportcasters, had told me of attending a Western media briefing in Asheville at which the attendees attacked the fare as it was served. Waters sat by helplessly. After a few minutes, the coach quietly asked the waitress if she'd please cut his meat and feed him a couple of bites.

"Talk about feeling small!" said Brown.

Well, they've named the playing surface over in Cullowhee "Bob Waters Field" now. Western still has to schedule the football superpowers to make ends meet. And Hollywood has yet to make "Pride of the Catamounts."

But the Muscular Dystrophy Association has inaugurated a "Bob Waters Courage Award" for young athletes who demonstrate some of Bob's tremendous grace and bravery.

For that, MDA has my appreciation and my check.

Spring 1991

Ed White

Ed White of Davidson (R) won plaudits as his conference's leading scorer in basketball. A scrappy football halfback, he became a coach, administrator, and a seniors tennis champ.

At halftime, the first year coach's team trailed the state champion Yellow Jackets, 34 to zip.

After haranguing his shell-shocked players, drenched by the steady downpour outside, the young coach approaches his smallest and least talented team member.

"Jake Cheatham here could have played better than any one of you out there," he said, laying a hand on my padded shoulder. "Jake Cheatham will start the second half and play every minute of it. One thing he's not is a quitter!"

Well, that was Ed White, the Southern Conference's leading scorer in basketball at little Davidson College the year before. And at 5 foot, 8 inches and 142 pounds, a letterman in football.

This tough-as-a-pine-knot competitor presented me the only athletic letter I would ever win. And I could identify with this intense little man with the big heart. Maybe that's why I let him call me "Jake" even though my name is Mike. I just respected him so that I thought it impertinent to correct him.

H. Edmunds White, the China-born son of Presbyterian missionaries, later became director of admissions at his alma mater. He specialized in recruiting future Rhodes Scholars. Such was his interest in athletics, though, that some of them could score 20 points a game or run a 4.5 in the 40-yard dash.

Then White became athletic director at Rhodes University in Memphis, another of those academic strongholds that feature athletes who play strictly for the love of competition. And somewhere along the way, he took up tennis. With a passion. (Anyone who knows him will tell you he hadn't the patience for golf.)

Now in his mid-sixties, my old coach is annually ranked in the top ten in seniors play in the Atlantic states.

"Ed White is the most awesome competitor I've ever known," testifies Ed Farrell, former athletic director at UNC-Asheville and football coach at Davidson.

Now, about that second half I started in the downpour. Well, I played all right and the champs only scored two more times.

My teammates told me, "Jake," — the scrubbiest of the scrubs — that I showed coach what heart really was.

The other day, Coach . . . still possessed of all his intensity, all that integrity . . . called. Said I did more with less than any of his players.

Then he asked, "Say, what was the name of that boy who ended up as a starter at . . . ?" He called the name of a school that plays big-time football.

Mr. White is still coaching. This time the team is a coed junior high school team around the Davidson area. Now that's heart! And that's the fella who remembered ol' "Jake."

And to my knowledge, he's never called the boy who went on to play big-time ball. The one who quit on him that night when the bottom fell out of the sky.

Oh, that day at assembly when Coach awarded me the letter, the coach in my family said, "Son, we'll go to see the Bulldogs Saturday, and I'm gonna get you the finest lookin' white sweater they got at McGregor's to wear with that big purple 'D.' "

And you know, if Coach Ed White comes up and puts his hand on my shoulder at halftime some night, I won't let him down then, either.

Winter 1991

Junior Wooten

Time was in the late 1940's when life was simple. And we were at peace. But some veterans still wore their "Ruptured Duck" discharge pins.

In those happy days, my dad used to take me out to the old ball park on Sundays for a double-header featuring a couple of teams from the long-defunct Northwest Georgia Textile League.

Or maybe all the way to Atlanta in our 1940 Chevrolet, which Dad refused to drive at a speed over 40 miles per hour. Ah, to Atlanta and Ponce de Leon Park, palace of the AA Atlanta Crackers!

You played for Atlanta, it seemed, and you had an automatic ticket to the big leagues. If that's what you wanted.

But a lot of those boys from the mill towns in Georgia and the Carolinas were content to stay home and play for the mill team. Right smart of 'em did venture out to perform for Atlanta or the Charlotte Hornets or the Asheville Tourists. For 12 or 14 years maybe.

A few of them played for the mill basketball team in the off-season. One of those was my boyhood idol, Earl (Junior) Wooten of the Crackers and South Carolina's Pelzer Mills.

Ah, Junie! He never seemed to hit below .300. And he scored 40 or 50 points an outing for the mill basketball team.

Once someone persuaded Wooten to move up to the Washington Senators for a couple of seasons. And was he ever homesick! He just had to hear the old mill whistle.

So, Junie hopped a Greyhound and came home to Pelzer.

There were dozens like him in those mill towns, and could they play baseball! Cliffside and Avondale had teams. So did the Enka plant, near Asheville. They all did.

Other than little Junie Wooten, my heroes were "Bad Eye" Lindsey and Lefty Hutchins of the Northwest Georgia Textile League. They ended up playing for almost every team in the league, except maybe Aragon. That club, they were real losers.

They wouldn't even put up the money to lure away a great draw like "Bad Eye!" What kind of mill was that?

About my dad, he was an old semi-pro catcher. When I was little and he'd hold my hand as we walked to the ball park, I'd try to massage those gnarled old hands of his, broken countless times by fouled-off pitches. (Forest City's Smoky Burgess has hands like that.)

"Don't worry, Liddle 'Un," he'd say, "they don't hurt anymore."

We'd eat hot, salty goobers and drink ice cold Coca-Cola from the town's bottler, Mister Willie Barron. Mister Willie's franchise had the highest per capita consumption mark in the world. Really! And Dad and I helped keep him there! Least we could do for a nice man like Mister Willie!

Then we'd come home and compliment Mama on her spoon bread at Sunday night supper. And then we'd set down and listen to the Jack Benny Show, the whole family would.

That night I'd go to bed to dream about growing up to be just like Junie Wooten.

I called information in Pelzer the other day. There's no Pelzer Mills in the book now. And no Earl Junior Wooten either.

But I remember Junie. Just like it was . . . only yesterday.

Winter 1991

Dyan Zumstein

"Welcome to The Center of the Universe," the program announcement read. It was a local bank's way of expressing civic pride in the fact that the Southern Conference Basketball Tournament is held annually in its town.

"I mean, really," says Dyan Zumstein, a cheerleader for the Western Carolina University Catamounts, "they don't come any bigger than this one!" And, in fact, folks associated with the venerable conference's member institutions tend to share Miss Zumstein's rather high opinion of the event. A tourism promotion group has even rated it among the top twenty annual attractions in the Southeast.

Look out, Mardi Gras and Spoleto Festival!

Residents of Asheville, North Carolina (population 61,607), appreciate another aspect of the three-day gala here. It generates about $2.5 million for the local economy. And the folks hereabouts, it seems, will go to any length to see that the Southern stays.

This, despite annual efforts by the likes of the Charlestons and the Winston-Salems of this basketball-crazy region to wrest the jewel from its "Land of the Sky" moorings. (The tourney has been held in Asheville for nine consecutive seasons.)

I'm due for a reality check now, you suggest? This isn't Madison Square Garden, the NCAA, or the ACC shootout. Right?

Right, and the eight-team conference — soon to feature

Georgia Southern and bring back Davidson — according to *USA Today* has a so-called power rating for the cage game that places it a mere twenty-third among the nation's college athletic loops. Still, on an overall basis, a good many observers consider the SC the finest classification I-AA conference in the nation.

So what if they're a notch below the big boys competing in I-A?

Affable Wright Waters, the Southern's commissioner, knows why his league enjoys such high regard. "Our real strength is derived from the member institutions — along with their coaches, athletes, and supporters." (The members are: Appalachian, East Tennessee, and UT-Chattanooga; also Furman, Western, and Marshall; and finally the military colleges, VMI and The Citadel.)

Among the supporters — and concerning some long-running natural rivalries — consider the aforementioned Dyan Zumstein. Ask the lithe and athletic WCU junior cheerleader about her 'Cats and the Mountaineers of Appalachian State University. Her expressive blue eyes glimmer like mountain foxfire when she replies, "A win over the Apps is always a VERY big one."

True. In basketball, the two rivals have met 131 times, with ASU recording 81 victories. When both were "normal schools" they met for the first time at Boone during the 1927-1928 season.

In fact, the contest had to be halted as officials cleared away debris thrown onto the outdoor court. The Apps' fans had chosen to demonstrate their displeasure over a ref's call in rather dramatic fashion.

No such antics at this year's three tussles. But it's for certain that not only the athletes are into the rivalry. Zumstein, for instance, is one of a dozen highly-skilled and committed support people. The cheerleaders are individuals with solid credentials in gymnastics, dance, and a variety of athletic disciplines.

They have their own coach, as well. She is Sue Collins,

a local lady who volunteers time after work each day as a speech therapist.

About the comely Dyan Zumstein, does she aspire to become, say, a Dallas Cowboys cheerleader? And does the NFL conduct a draft for cheerleaders? Negative on both counts, of course.

This twenty-year-old from Etowah, near Hendersonville, has done some serious career planning. A graphic arts major, she plans to latch on at one of those Disney theme parks, perhaps even as a Mouseketeer. Once there, she expects to move into the animation end of the vast combine's film properties.

A championship clogger who performed at the 1984 World's Fair and who has appeared on the Nashville Network, Dyan has been a dancer since age four. So she has more than one string to her bow.

Smart, very smart. (And she won her school's coveted "spirit award" to boot for 1991-1992.)

And, oh, while the Catamounts were eliminated by ASU in the tournament's opening round, one expert on cheerleading tells me she and her colleagues were tops among those competing at courtside.

Other heroes were the students and fans who jammed Asheville's downtown for three days. A local policewoman, Connie Robinson, tells me only two were advised to head for the city limits. There were no arrests.

Contrast that record with the annual State Fair weekend in Dallas where football squads representing the Universities of Oklahoma and Texas square off in the Cotton Bowl. While living in the Texas city I never recalled a single year when less than 300 were rounded up and hustled to the calaboose.

Pretty high type folks at our Southern Conference schools. My friend, restaurant man John Shubash, agrees. "They seemed real happy to be here, and we were mighty glad to have 'em here; good tippers, too!"

Who knows, it just may be that in the NCAA tournament, the champions from ETSU defeat one of those high-and-mighty division I-A teams this go-'round. (And that they did, taking the measure of the PAC-10's Arizona in the opening round before falling to tradition-rich Michigan from the Big Ten conference.) But their fans probably won't have as much fun at that tournament as they did up here in these mountains of ours.

Hey, for three fantastic days we WERE at The Center of the Universe, weren't we?

Well, it did seem that way to me and Dyan Zumstein. And to Mark Wilson, the ad agency fellow who wrote that copy for the First Commercial Bank.

Man, oh man, "Center of the Universe," that's us 'uns! "Whoo-boy!" as comic strip character Pogo used to exclaim. We got ourselves quite a tournament up here!

(Or was that Lou Bissette, our town's former mayor and quintessential sports fan, who did that "Whoo-boy" number?)

Winter 1992

Dyan Zumstein is a Western Carolina University cheerleader who is the apotheosis of school spirit at events like the Southern Conference Basketball Tournament, played in Asheville, North Carolina, in 1992 for the ninth consecutive year. Dyan reminds us there's much more to athletic endeavor than just the athletes competing.

. . . And Sundry Other Heroes

Red Barber

Red Barber. is as Southern as the old mill whistle at one of our many textile plants in the Carolinas. And many of his listeners regulate their lives around his Friday broadcasts, much as workers at the now-disappearing mills lived by and for the old whistle.

His "Barberisms" during hundreds of sportscasts over the last six or so decades have endeared him to us, almost as though he were an especially cherished neighbor. Or maybe your wife's good-hearted uncle visiting the house on a wintertime holiday. An uncle, say, who lingers over a second cup of coffee to share a few of his vintage yarns and maybe leaves off an autographed photo of a baseball star for a flu-ridden Sammy.

The Mississippi-born national treasure was christened Walter Barber 84 years ago next month. And he is, by any criterion I personally apply, simply the warmest and most refreshing commentator in the media today.

I wonder to myself, is it because he's another "Son of the South," much like the men I grew up idolizing? Men like our family physician, Dr. John H. Terrell; our minister, Dr. A.T. Cline; and my own father.

This I do know: Red loves his bride of 61 years, Lylah — and the camellias he grows in the backyard of his Tallahassee home. Maybe more than baseball itself. (He once epitomized changes in the game with this observation: "When I was in baseball and you went into the clubhouse, you didn't see ballplayers with curling irons.")

Another thing I know just as surely as the fact I simply cannot pass up blackeyed peas and potlicker, this man is a believer. He once made that clear in a book he wrote on Christians in the world of sports, *Walking in the Faith*.

What I asked Red Barber recently, though, is why it is Southern sportscasters have such high levels of acceptance among their audiences. I cited the examples of people like Barber's former colleagues, Mel Allen, Ernie Harwell and Keith Jackson, as well as that of The Redhead himself.

It would be irresponsible, he said, to assume "we were a preponderance" of the successful sports voices. It's more of "an individual thing," he added.

Even so, I do think the unique manner of speaking of our native son sportscasters have set them apart. Their sayings are astonishingly — and illuminatingly — appealing to fans from other regions.

What great fun it was to enjoy Red during his accounts of the games of the Dodgers, the Yankees, and the Reds when he would pipe up with:

- "He's sittin' in the catbird seat!" (Pitcher is sitting on a lead.)
- "Oh, Doctor!" (A sudden turn of events.)
- "Hold the phone!" (The manager is going to the mound.)
- "They're tearin' up the pea patch!" (His Dodgers are scoring.)

"This ol' country boy" (as he often describes himself) was there when almost everything of historic significance transpired — or so it seems. For instance, the very first words spoken at the debut television broadcast of a major league baseball game were, "This is Red Barber speaking. Let me say hello to y'all."

Recently *Broadcasting* Magazine selected, on the 60th anniversary of its founding, 60 individuals who affected its industry most. Of course, Barber's name is there. And he's still at the mike, offering kindly commentary and gentle wisdom on sport — and life in general — on National Public Radio.

You'll become aware that this institution in our midst is one who lives his personal statement of faith. It's much like this one: "Remember yesterday; dream of tomorrow; and live for today."

Happy 84th birthday, Red Barber, on Monday, February 17. I'll see you on the radio Friday morning, okay?

Winter 1992

John Coke Brown

John Coke Brown of Georgia took as his middle name the tradename of the product he purveyed. Receiving All-America mention in football, Brown also was named to *Sports Illustrated's* Silver Anniversary Team and won the Distinguished Service Cross for bravery in World War II.

Growing up in Toccoa in the 1930s and '40s, I remember my dad's observation: "Play football for Georgia, son, and you'll get a job with Coca-Cola!"

That sounded good to me, and in fact, in 1959 I did land a job with that fine home state company. But another long-cherished dream — playing ball for the Bulldogs — collided with the stuff of hard, cold reality when the best a 129-pound halfback could manage was three minutes playing time for the 1948 Darlington School Tigers in Rome.

When I came on at The Coca-Cola Company, Georgia had its own roster of UGA grid stars, several in the executive ranks. They were scattered throughout the company's farflung empire, fierce competitors all in the soft drink wars, as they had been in their playing days. The opponent this time, however, hailed from a place other than Atlanta or Knoxville or New Haven, Connecticut.

(And, you know, I've been gone from that company 25 years now, but I still just can't bring myself to say "The 'P' word," much less drink the stuff. Jim Wimberly, another UGA product and a Coke vice president, never said that word either — just called it "PC," and he wasn't using the familiar South Carolina acronym for Presbyterian College.)

Over the years at Coke, there were Joe Bennett, All-America at tackle in 1924; J.C. Hall, a standout guard in 1936; and end Herb Maffett, a college All-America in 1930 who played at Toccoa High School. There were others: "Daredevil Dave" Collings, 1919-1922; Jim Carmichael, an end in the '30s; "Tiger" Bennett, a guard, 1931; and Buck Chapman, fullback, 1932 through 1934.

John Brown of Athens gave the Dogs three fine years as well, 1932-1934. He was honored as most valuable player his senior season by a vote of teammates and won all-conference honors, as well as All-America mention.

The Athens High product ended up running company-owned Coke plants in Grand Rapids and Lansing, Michigan; in Seattle, San Francisco, and Columbus, Ohio. And it was while serving as president of The Coca-Cola Bottling Company of Ohio that the slow-talking, analytical executive was named a *Sports Illustrated* Silver Anniversary All-America selection.

A Silver Anniversary All-America? Yessir, for the years 1956-1966 the publication honored former college gridders on the 25th anniversary of their senior seasons. Selection was made on the basis of achievement in their chosen fields of endeavor — law, business, politics, medicine, coaching and the like.

Coach Woody Hayes of Ohio State, to become Brown's pal in Columbus, also was named that year, and together they flew to New York for the induction ceremony. Another honoree was then U.S. Representative Gerald R. Ford, later to become the 38th President of the United States.

The Georgia man among them — always one to seize upon opportunities to plug his product — was photographed belting down some "Atlanta holy water" while standing beside one of his trademarked drink coolers. He stood ramrod straight, this former UGA ROTC grad, in shirt sleeves, his steel gray crewcut, as always, at attention.

"The Big Beverage," by E.J. Kahn of *The New Yorker,* came out about this time and profiled John C. Brown. The Coke bottler in Columbus, Kahn discoursed, even listed himself in the telephone book as "John Coke Brown," thus distinguishing himself from the three dozen or so other John Browns in town. (He was christened "John Cook Brown," the Cook for his mother's maiden name.)

In so declaring, John left little doubt as to his loyalties. When I mention that passage in the book, John's rejoinder is, "Uh huh, and named my firstborn 'Georgia.' "

When we worked for this UGA loyalist in the Southwest Region of the company in the '60s, we all called him "Coach." But not to his face. (Van Chavis, a one-time college fullback, leaned over to me once at a meeting and whispered, "You see the forearms on that son-of-a-gun!")

The man who inspired tremendous loyalty among the troops could also instill the fear of God in young district reps, as we were called.

I often go to John and Adell's place in Bellevue, a Seattle suburb, on fall weekends to see the games on the tube. John views the contests from his "desk," the dining room table. I think he's plotting strategy to capitalize on Clemson's overshift on the weak side.

The appellation "Coach" was by no means inappropriate. As a young cavalry officer in World War II, Brown had coached the 1941 Fort Riley team to the all-cavalry championship, defeating the Fort Bliss squad 30 to 6. His star quarterback, Bobby Nowell, had performed for the Wally Butts-coached UGA teams of 1939 and 1940.

A miffed 1st Cav general from Fort Bliss stalked out at the half, dispatching his aide to tender Brown an offer in the dressing room after the game: Accept transfer to Fort Bliss, coach my team next year and I'll approve transfers for any of the ball players you want to bring with you.

The UGA grad declined, instead requesting duty on the front lines. He became a tank battalion commander, winning the Distinguished Service Cross. Duty, loyalty — just a couple of Brown's hallmarks.

Other young coaches won their spurs with service teams in those years when Hitler and Tojo broke up Georgia's incomparable 1942 Rose Bowl aggregation. Launching long and successful careers were service team mentors Paul Bryant of Alabama and Wayne Woodrow Hayes of Ohio's little Denison College.

Denison, by the way, I tell John, is the only college team in America still running the single wing. He chuckles and comments that when he coached Fort Riley he sure coached it, that the Notre Dame Box version of it was all he knew.

About the early days, he says he's always remembered the home officials at those little mountain towns' high school games back home. "At Cornelia in my first season at Athens High, I was trailing on an end-around play; runner fumbles, I pick it up and dash into the end zone." His voice rising, he exclaims, "So the ref calls it back and penalizes us for 'an intentional fumble!' "

About this time — and possibly to defuse John's indignation over that injustice — Adell calls, " 'Bud,' you and Mike come on to lunch; used my Swan House recipe for the chicken salad!" (I remember Winnie Butts' fried chicken; it was pretty remarkable, too.) Then I tactfully reroute the conversation to baseball and the Georgia Peach, Ty Cobb.

I tell Coach Wally Butts' story about going to an NFL game with an aging Cobb, who proceeded to start three fights. "Oh Lord, never again will I go to a game with that fellow," exclaimed the man we used to call "The Passmaster."

"Uh oh, here we are back to Georgia football again, 'Del," I apologize to her, " 'Scuse me, Ma'am."

I tell John I'm gonna write this piece. He growls, "Hey, boy, nobody wants to hear about me!" I persist. Later I call 'Del to persuade her to persuade John to let me do it. She does.

My old boss always got a kick out of my questions, back during our respective hitches in the Southwest Region, the ones about the Bulldogs of the Harry Mehre era. Still does, now that I live here in Seattle, where John and his wife, Adell (the former Adell Helmer of Atlanta) retired in 1977. John's last assignment was as chairman of the Pacific Coca-Cola Bottling Company here.

We talk about John Bond and John McKnight, the Toccoa twosome who cocaptained the 1935 Bulldog team. My dad was always very proud of the pair, seeing as how he had been superintendent of city schools and had done some coaching during his 19 years there. Joining the duo as a leader was John's close friend, Toccoa halfback Marion Gaston, another of my dad's boys. Gaston, like Brown, joined Coca-Cola after graduation.

John Brown remembers with a certain reverence Harold Hirsch, the late general counsel of Coca-Cola and arguably the leading UGA booster of his day. John's football scholarship came from an NCAA-legal scholarship from the "Harold Hirsch Fund."

He traveled to Atlanta from Athens and thanked the Great Man for his scholarship. Impressed, the grey eminence of Coca-Cola said, "All right, John, come to see me again when you graduate."

Never a man to forget a kindness shown him — he had played only one season and part of another of football at AHS; there were doubts he could play at the level of the old Southern Conference — John showed up four years later at Mr. Hirsch's law offices. Telling him he was responding to Hirsch's "come see me" invitation, Brown was told: "Okay, Johnny, you've done me proud. Now, I've got a couple of good clients where I'd like you to work, Scripto and Coca-Cola. Which would you prefer to work for?"

The young AHS graduate 'lowed as how he reckoned it'd be "Co'-Cola." That scholarship — and the Hirsch "invitation" — brought a real talent to that marketing marvel of a company.

Years ago, I reminisced with John Coke Brown, telling him Dad commented about my new boss, "Sure, I remember him, scrappy but kinda light." John loved that, for he was a muscular 230 at the time. His playing weight had been 185 as a pulling guard in the Notre Dame Box formation Mehre brought with him from South Bend.

Number 17, John Brown of Athens, these days is working his way back to his Coca-Cola playing weight of 230. He has a collapsed diaphragm and must wear an oxygen canister at all times.

At 77, his memories of Sanford Stadium are as sharp as his football and marketing acumen. And he tells of playing before 100,000 at Yankee Stadium in New York against Eastern power NYU. This, while two other Eastern powers, Fordham and Army, drew 40,000 across town on the same Saturday at the Polo Grounds.

He regrets Georgia's having been able to schedule another superpower, Yale, only sparingly during his playing career. The Elis, Brown says, were embarrassed by defeats in earlier years at the hands of the upstart Southern institution. They wanted no part of the Dogs.

One of Brown's best UGA/Coca-Cola friends — probably

his best — is one-time boss of company-owned bottling operations, Morton S. Hodgson, Jr. He, too, was an Athenian.

The former Coke official now resides in Highlands, North Carolina, and the two are often in touch. John succeeded Hodgson as a corporate vice president.

Hodgson's brother, Hutch, incidentally, was a Bulldog guard in Brown's time at the university. Hutch, now deceased, is the father of Pat Hodgson, a fine Georgia end and later offensive coordinator of the New York Giants.

John Brown, you should know, lost his physician father, John Andrew Brown, at age eight. Months later his mother died. Three children left Sparta in Hancock County. John and brother Robert moved to Athens to live with an uncle, Dr. Wedford Brown, a teacher and coach at Athens High. Sister Electa went to live with grandparents. Robert still lives in Athens, where he collects UGA information from Coach Bill Hartman to relay to his brother in Seattle.

"I used to love making house calls in Hancock County with my father; oh, how I loved doing that," said John. "I think my dad and my uncle must have been a lot like your dad. . . ." His voice trails off.

Yes, John, I think so. They all three had a great way about developing young men. "Coach" had that ability, as well. He led and inspired dozens of future top managers and executives for the soft drink superpower.

John Brown and his Bulldogs, John Brown and his Coca-Cola: Two great passions in a life worthy of emulation by today's graduates.

Oh, yes. If Georgia should score the winning touchdown to down Auburn for the SEC Championship this fall, hoist a Coke to your lips and savor the moment.

John Brown will approve. A Coke man, he's a team player — and a Georgia gentleman who's done right well by the two institutions he believes in most.

He's the real thing: a Coke Classic, a UGA Classic.

Summer 1990

HEADQUARTERS RESERVE COMMAND
7TH ARMORED DIVISION

A.P.O. 257, U.S. ARMY
29 June 1945

201-Brown, John C. (Off)

SUBJECT: Recommendation for Award

TO: Commanding General, Seventh U.S. Army, A.P.O. 758, U.S. Army (Thru Channels)

1. Under the provisions of AR 600-45, it is recommended that **JOHN C. BROWN, 0 315 210, Lt. Col., Cavalry,** be awarded the **Distinguished Service Cross,** for **extraordinary heroism in action.**

2. a. Lt. Col. BROWN was serving as Battalion Commander, 40th Tank Battalion at the time of rendition by him of the service upon which the recommendation is based. Lt. Colonel and assigned to 40th Tank Battalion at the time service was rendered and at the time recommendation is made.

b. Previous awards:

Awarded Silver Star per GO 42, Hq 7th Armd Div., 8 September 1944.

Awarded Oak Leaf Cluster to Silver Star per GO 64, Hq 7th Armd Div. 13 October 1944.

Awarded Purple Heart per GO 2, Hq 7th Armd Div. 1 January 1945.

Awarded Oak Leaf Cluster to Purple Heart PER GO 61, Hq 7th Armd Div. 10 April 1945.

Awarded Oak Leaf Cluster to Purple Heart per GO 62, Hq 7th Armd Div. 11 April 1945.

Awarded Bronze Star Medal per GO 31, Hq 84th Inf Div., 19 February 1945.

Awarded Oak Leaf Cluster to Bronze Star Medal per GO 45, Hq 7th Armd Div., 27 February 1945.

A similar recommendation has not been previously submitted.

c. The entire service of Lt. Col. Brown has been honorable since the rendition by him of the act or service upon which this recommendation is based.

d. The officer recommending this award was eyewitness to the act or service upon which this recommendation was based.

3. a. Complete description of action or service: For distinguishing himself by extraordinary heroism in connection with military operations against an armed enemy of the United States in Germany on 26 March 1945.

Poised on a perimeter of the Pemagen Bridgehead early on the morning of 26 March 1945, Task Force Brown with Combat Command Reserve, 7th Armored Division was in a position to make the principle effort of the First Army to break out of the Bridgehead and make a deep penetration into enemy territory. The assigned mission was to secure an objective approximately 40 miles distant, bypassing resistance and disregarding the taking of prisoners if necessary. Upon encountering stiff enemy resistance and impassable obstacles at DIERDORF, Germany, Lt. Colonel BROWN went to the forward elements of his command in a 1/4-ton truck where he immediately came under direct anti-tank fire. In this action his vehicle was wrecked and he suffered a painful leg injury. Refusing medical aid, he remained in this position which was subjected to heavy artillery and anti-tank fire and directed the two sub-elements of his Task Force in a double envelopment of the enemy position. With this operation successfully under way he returned to the aid station for treatment. He again refused a medical officers' recommendation that he be evacuated, and securing another vehicle he immediately returned to the head of his command and succeeded in overrunning the enemy position. He then directed the reorganization of his forces and began the advance to his objective. Realizing that the advance was slow and hesitant, Lt. Colonel BROWN moved to the head of the column in his 1/4-ton truck at great risk to his own life urged his forces forward in a bold and aggressive manner. Upon overtaking a retreating enemy column he mounted the rear deck of the leading tank and directed the operation from this exposed position, disregarding enemy small arms fire and returning fire with a sub-machine gun and pistol. By his audacity, boldness and utter disregard for his own safety he so inspired his men that they swept forward with all possible speed and fire power, destroying column after column of the retreating enemy. Vehicles and equipment caught in the path of the advancing tanks were literally crushed before they could resist or escape. Enemy tanks, 88MM anti-tank guns, and artillery pieces were destroyed before they could be placed in effective firing position, enemy infantry could deliver only sporadic fire, so vicious was the attack. For four hours the attack continued. Lt. Colonel BROWN was always in an exposed position with the leading elements, frequently the target of enemy small arms and bazooka fire, continually seeking the position from which he could best direct the attack. At one point an enemy bazooka struck the tank upon which he was riding and he miraculously escaped injury, yet his courage and determination remained undaunted. Upon nearing the objective at HECKHOLZHAUSEN, Germany, the task force was so deep into enemy territory and so far ahead of adjacent friendly forces, that our planes refused to believe it to be friendly and harrassed the column by straffing and bombing. Although Lt. Colonel BROWN narrowly missed death from an exploding bomb, he continued to push the attack with more and more vigor. At the close of the operation an estimated 10,000 German troops had been neutralized and started to the rear. Three enemy tanks, seven self-propelled guns, eleven medium artillery pieces, fifteen heavy artillery pieces, eleven 88MM dual purpose guns, one hundred motor transport, and over two hundred horse-drawn vehicles had been destroyed or captured. Approximately two hundred enemy were killed and three hundred fifty wounded. The unparalleled success of this operation was a vital factor in the breakthrough operations of the First United States

Army and the ultimate encirclement and destruction of The German Army in the RUHR pocket. The extraordinary heroism and the gallant leadership displayed by Lt. Colonel BROWN was the impetus and the inspiration which led his small force of 600 men to such an accomplishment.

His actions demonstrated an intrepid devotion to duty, and exemplify the highest traditions of the armed forces.

b. (1) Terrain rolling; hostile observation excellent; enemy fire sporadic and light.

(2) Visibility very good; time of day 1145 to 1545; atmospheric conditions cloudy to clear.

(3) Location of enemy — all along axis of advance, dug in, in prepared positions, in assembly and bivouac areas.

(4) Morals of our troops superior, enemy troops initially excellent.

(5) Casualties sustained — own negligible, enemy approximately 10,000 in killed, wounded, captured plus enormous materiel stocks.

(6) Effect or results of deed cited — a very vital contributing factor to the breakthrough operations of the Remagen bridgehead which ultimately led to the encirclement and destruction of the German Army in the Ruhr pocket.

4. a. JOHN C. BROWN, 0 315 210, Lt. Col., Cav., 40th Tank Battalion, 7th Armored Division, for extraordinary heroism in action against an armed enemy of the United States, on 26 March 1945, in the Dierdorf-Heckholzhausen, Germany, area.

b. Encountering stiff enemy resistance and impassable obstacles at Dierdorf, Lt. Col. BROWN advanced to the forward elements of this Task Force in a 1/4-ton truck to actively direct his troops. He was immediately subjected to direct anti-tank and artillery fire, his vehicle destroyed, and he painfully injured. Refusing medical aid he directed and led his troops in a magnificent and successful movement. Throughout the operation he remained at the head of his column, rotating between 1/4-ton truck and tank as occasions demanded. He rode on the rear deck of a tank, exposed to enemy fire and wrecked havoc on withdrawing German troops with his own weapons.

c. The skill, daring, and utter disregard for his own safety were an inspiration and stimulating influence to officers and men alike; contributing immeasurable to the success of the operation; and are in keeping with the highest military traditions. Entered the military service from the State of Ohio.

s/ Francis P. Tompkins
t/ Francis P. Tompkins
Colonel, Cavalry
Commanding.

A TRUE COPY:

Signed
CARL E. BOSECKER
Major, Q.M.C.
Actg. Asst. Adj. Gen.

Jack Cannon

"Black Jack" Cannon, consensus All-American at guard for the 1929 Notre Dame national championship team, also was named to the collegiate Football Hall of Fame. Cannon's stern, football-hating father became a fan — of sorts — after Cannon's performance in a game at South Bend.

Eleven-year-old Annie Cannon wondered what the big deal was early that fall Saturday morning in 1927. The extent of her knowledge about the 232-mile journey that lay ahead was that her family was to visit Brother Jack at the big university in South Bend, Indiana.

(The account that follows may be at odds with the actual event as it in fact transpired. But I think it may square with major aspects of the sojourn as recalled by Annie some 50 years after the family's trip that fateful Saturday. I hope so, for to me the Cannon family's experience that day definitely is one to be shared.)

The Cannon family of Columbus, Ohio, loaded five of its members into the Studebaker sedan at 3:00 a.m. Sleep still crowded the children's eyes. Each had a sack lunch, and there was a final admonition to go to the bathroom before departure. "Can't be stopping every half hour, you know," Father said.

The three girls wore the dresses they usually reserved for mass on Sundays at the Cathedral of St. Joseph's. And Father had brought Mother a new handbag from his dry goods store for the occasion.

Annie noticed that Father wore his three-piece business suit with the watch chain across the vest, along with his usual derby. But for this special occasion, he carried a jaunty new walking cane. One with a shiny silver ornament perched atop the stem of the gutta percha cane.

To the young girl, it seemed a whole lot of trouble to go to, visiting Brother Jack — who, for Annie's taste, thought himself far too handsome for his own good.

The youngster also was aware their father was less than enchanted that Jack had chosen to go out for the football team at the college, Notre Dame. "Not a game for gentlemen," Father had muttered. The stern pater familias was very keen on clean living and of course abstention from "demon rum."

Annie fell asleep again, and once awakened by her sister Helen at daybreak the girls commenced a game that involved counting red barns along the roadside. "Not fair, Helen! That one wasn't even red!"

Father eventually would lead his entourage into cavernous Cartier Field. "At left guard, number 40, Cannon," entoned someone using a megaphone. "Fight, fight, Cannon!" came the cheer. Father's jaw tightened. But, curiously enough, as the game progressed Father grew somewhat animated. Even smiled a bit when a man three rows down piped up, "Stout fellow, that Cannon!"

Toward the end of the contest, the opposing team fumbled. Brother Jack scooped up the pigskin and sprinted into the end zone for the winning score. (And, yes, "Black Jack" Cannon would become a consensus All-American in 1929 as the Irish won the national championship. And in 1965 he would be the first of six Notre Dame guards to be voted into the Collegiate Football Hall of Fame over the years.)

Ah, yes, after the cheering subsided that afternoon, Mr. Cannon gently tapped a gentleman on the shoulder as he resumed his seat on the row below. "Pardon me, sir, my name is Cannon, and I'm from Columbus, Ohio." Detaching the shiny ornament from the cane, Father continued with a polite smile upon his face, "Will you join me in a drink, sir? That young man is. . . ." His voice broke and then he managed to say, "That young man is my son."

"Black Jack" Cannon's flamboyant style of play came to be his hallmark, it can be added. Writer Dave Condon, in a history of Irish football, says his single most vivid recollection of the team's championship season was of Jack Cannon charging down under kickoffs without benefit of headgear, his black mane waving in the wind. (The practice was outlawed in the college game the following season.)

Well, about that memorable day at old Cartier Field, that's the way I imagined it from the accounts of Mrs. Annie Gugle.

What a day, what a game, and what a brother she must have had. But unlike writer Dave Condon, whose fondest memory was of the flying black mane of the hatless Jack Cannon, it was a different kind of memory to the Cannon children. Whatever else it was, to the Cannon family it was, quite simply, "Father's Day."

Spring 1992

Leo Costa

The proud grandfather is J.B. Ryner, an old schoolmate of mine. I have known J.B. and his wife, the former Anita Sams, for 40 years now. And the grandson's name is Derrick Ryner.

The card Ryner passes around at the pre-game tailgate party is modeled after a sales order form. (A good salesman J.B. is; after all, he got Anita to marry him.) The sales order form lists several options for the "buyer." A couple of them:

– "Pictures of Grandpa lifting baby grandson."
– "Pictures of baby grandson lifting grandpa."
– "All of the above now available; order early, avoid the rush."

My old pal tells me, "Is that boy big? Lordamercy, yes." Then concern etched on his face, he adds, " 'Course, if he can't be a fullback – if he's just too small – I'd like him to be a kicker!"

Now, at the same gathering of old grads stands Leo Costa, the school's most-successful-ever booter at the time of his 1942 team's big win over UCLA in the Rose Bowl. Leo, he just chuckles at ol' J.B.'s suggestion that the little fellow "can always be a kicker."

The former kicker's attractive other half, Carolyn, then points out, "Beginning this year, Leo says those kickers are going to have to play it honest like he did." Carolyn's right. This season the rules-makers have decreed that the uprights must be narrowed from 23 feet, four inches, to 18 feet, six inches — same as in the National Football League.

In this year's opening game on August 28, both Penn State and Georgia Tech even missed a point-after try. The next night Florida State missed one of those "gimmes," as well, against Brigham Young University. Only one field goal was kicked in both games.

Moreover, while three-pointers accounted for 13 percent of total scoring over the course of the 1990 college campaign, look for that to change.

Yeah, look for that to drop to maybe three percent this time out. And look for the coaches to recruit kickers who can play an additional position. Like Leo Costa did; he was also a backup guard.

One of those college mentors made a distinction worth recording. "In the NFL if a guy misses too many of those kicks, they fire the kicker. We can't fire the kicker," the coach points out. And coaches in other conferences are probably of like mind.

In 1990 Southern Conference booters were a combined 11 of 15 (73.3 percent) on field goal attempts through the first two weeks of the season. This year, at the same juncture, they are only 7 of 18 (38.9 percent).

Well, about the time young Derrick Ryner is ready to walk on at the old school — I believe that'll be about the year 2006 — maybe steroid use will have ceased.

Or the little guy may look at much-larger teammates and change his sport. (Maybe tennis, J.B.?) Anyhow the proud grandpa says the projected fullback won't really have to have a football grant-in-aid to get through the university. Gramps already has opened a savings account to cover the costs of Derrick's college education.

Now that strikes me as a pretty good way of insuring the young man gets some learning. And it sure beats a four-year ordeal of trying to get the ball through narrowed uprights Saturday after fall Saturday.

Summer 1992

Leo Costa was prolific as a kicker for Georgia's bowl teams in 1941 and 1942. He competed at a time when the uprights were narrower and kickers were expected also to play a given position in addition to specializing in PAT's and field goals.

Cleophus Goodgame and Friends

When you have a last name like mine — Cheatham — you become more conscious of others' names, the given ones and the surnames. And you develop coping behaviors to endure the not-so-funny jokes that some folks make about your particular handle.

My maiden aunt, Mont, who died at 102 last May, had a very effective means of coping. Some wiseacre would ask, "Oh, you cheat 'em?" Then she was ready with this soft-spoken retort: "Well, you have to understand that's my name, not my intention!"

Gomer Hodge of Rutherford County once lamented with me about his nickname. Seems his Cleveland Indians teammates stuck him with "Gomer" (as in the television series character, Gomer Pyle). And the old infielder allowed as to how he probably did speak with a decidedly Southern accent, as did the television character.

Now, baseball always has been big on nicknames. For example, I'll mention "Joltin' Joe," "Leo the Lip," and "The Splendid Splinter." And you no doubt quickly will furnish the respective surnames Di Maggio, Durocher, and Williams (Ted, that is).

Many of you are so steeped in the lore of the diamond that you'll know too that "Old Aches and Pains" is Chicago White Sox favorite Luke Appling. True, a mere handful of us will identify "Three Finger" Brown as Mordecai Brown.

No such column is complete, of course, without mention of Frankie Frisch, "The Fordham Flash"; Connie Mack, "The Tall Tactician"; and Harry "Suitcase" Simpson." And toss in for good measure sundry "Schoolboys," "Duckys," and "Rippers."

Oh, the record books are replete with the "Peppers," "Kids," "Chiefs," and "Windys." Even a few "Deerfoots," and "Motormouths."

So far, so good, but let me make a few nominations for — Ta-da! — The Baseball Nickname Hall of Fame. Here they are: Louis "Buttercup" Dickerson, Harry "Tomato Face" Cullop, Dain "Ding-A-Ling" Clay, Joe "Goobers" Bratcher. Also consider for induction Anthony "Bunny" Brief, "Glass Arm Eddie" Brown, and Erv "Four Sack" Dusak.

I'll save Frank "Dodo" Bird for another day. But I am compelled now to share with you the absolutely perfect monicker for an athlete.

Honestly, I saw this one in the writeup of a prep basketball contest 39 years ago. The high scorer? None other than guard Cleophus Goodgame.

Have a nomination for the baseball nickname hall of fame? Drop it to me care of *The Star.*

Fall 1991

Woody Hayes

His gridiron program at Ohio State University was one mirroring the very best in the college game, they said. But there was a rising chorus of dissent about it as well, most of the time.

Coach Woody Hayes was a General George C. Patton of a college football coach who inspired intense loyalty among players and Ohio State fans. He had scholarly interests and a soft side few got to know.

You go to Columbus, Ohio, as I did several days ago, however, and you are struck again with the high esteem in which the late Woody Hayes is still held.

Walking into the massive St. John Arena at 410 Woody Hayes Drive, I passed a dominating photo of the old coach as I made my way to the school's sports information office. Why such worship, you ask, for a man who in his last game — the 1978 Gator

Bowl in Jacksonville — lost control and actually took several swings at a sub linebacker, Clemson's Charlie Bauman?

Whenever I mention his name down here, that single incident is all folks seem to remember about the man. And that's tragic, as sad as the inevitable firing that followed.

The record over 28 campaigns at OSU, with a single losing season: Four national championships, 13 Big 10 crowns; 205 wins, 61 losses, and 10 ties — a winning percentage of .771. This at an institution that boasts as rich a pigskin heritage as any of them.

The city where my 16-year-old daughter lives has a magnificent obsession — Ohio State Football . . . and its patron saint, the late Coach Wayne Woodrow Hayes. Ah, Woody, he was an unpredictable, stern and combative character in the mold of Duke's legendary taskmaster, Wallace Wade, and Georgia's tough-as-pigiron Wallace Butts.

With Hayes the game was all-out war. My old boss, war hero John Brown, told me no one did as good an imitation of General George C. Patton — in real life — as did Coach Hayes. Brown knew what he was talking about, he fought in the desert campaign in World War II and he and Woody traveled to New York to be inducted into *Sports Illustrated's* Silver Anniversary All-American Team.

This man Hayes, he never accepted a pay raise, refused to become athletic director (even though it would have meant more pay), and chose to remain with wife and family in the same modest bungalow on Cardiff Road they chose upon moving to town for the 1951 season.

Through all those 28 seasons, his wife, Anne, was the perfect counterweight to the coach. On game days, she sat intent upon the knitting in her lap as the game plan unfolded, never lifting her eyes to view the mayhem on the field.

One dear lady I know, related by marriage to the Hayes family, tells of joining the Hayes in their seats for a game one fine fall Saturday. After State scored its sixty-third point, she rose to make her way to the exit. "Sit down," Mrs. Hayes commanded, "We don't leave until the last gun sounds."

My friend dutifully sat down, and on that blustery day in March 1987 when Woody Hayes died she was among the many bringing flowers to the 50-yard line of historic Ohio Stadium in tribute to his memory.

My little girl was just three when the day after the slugging incident – at 6:45 a.m.; she was up watching the Saturday morning " 'toons." She came to our bedroom and exclaimed, "Daddy, Daddy, Woody Hayes has been shot and killed!" (She heard the word "fired" and to her it meant "shot and killed.") Even the toddlers knew and loved the man.

Dallas, my daughter, had met the living legend at an ancient Buckeye rite known as "The Senior Tackle." At that ceremony, those playing their last game for the Scarlot and Gray take one more crack at the tackling dummy. Woody shook the hand of each and told those assembled what each man had meant to the team.

Coach Hayes never lost touch with a player. They were his boys, and if they came onto hard times Woody was there to remind them they could win against the odds. A rugged old fullback, Jim Otis, said it this way when hearing of Hayes' death at 74: "He taught me a lot of football, obviously, but Woody Hayes will always be there with me in other things I encounter." Adding that he would never meet anyone of his stature, Otis said, "So many people in life try to play the part, Woody WAS the part."

The year after Hayes' firing, I spotted him in the faculty club one evening. (Mrs. Hayes was at his side, knitting.) He was preparing a speech, he said, and asked what I'd learned from playing the game that helped me in corporate life. I told him, and he asked about his old pal, my old boss John Brown. "Like I told you way back there, he was headed to the top," said Woody. Loyalty, nobody ever had more. A few unpredictable ways, maybe that too.

As I was leaving the building at 410 Woody Hayes Drive, I began to chuckle as I remembered the story about Woody and his next door neighbors, the Yuppies. Seems Woody on summer evenings slept out on the screened-in porch — in the nude.

Well, on this evening the Yuppies were dipping pretty deeply into their winecoolers. Their conversation went this way: "The old geezer is losing it, can't cut it any more!"

Then came a booming voice from the Hayes' porch, "Yeah, well, they say the old fool is deaf as a poker, too!"

Fall 1991

The Purple Hurricane

The mountains had names like Yonah and Currahee. The Tugalo River, six miles away, divided northeastern Georgia from South Carolina.

The name of the county seat, Toccoa, we were told, meant "beautiful" in Cherokee.

The town, part of what the government charts as the Chattahoochee National Forest, had a population in 1940 of 5,494.

And, the city high school with its 300 students was known for excellence in one-act plays and forensics at the district meets.

But, the consuming passion was football. It was self-expression, self-realization and a way of life. It was poetry, folklore and reality.

A fine one I am to recall the gridiron glories of the THS Purple Hurricane. My family moved when I was 11, so I never attended school there.

Having never worn the purple and gold, I can't summon up picture-perfect images of "the big game," an NGIC or NGFA title won – or an inspirational halftime talk from my late father, the school superintendent until 1943; or from coach Eddie Anderson, or from the team's great booster Dick Barron, whose brother David Irenus "Red" was All-America at Georgia Tech.

As I write this, it is mid-August in Columbus, Ohio, and the sports pages have started running their usual football roundup stories: "Miami U. picked to win MAC title,"

"'Buckeyes to Contest Michigan for Big 10 Leadership," and even here, 700 miles from Athens, "Georgia's Herschel Walker Cinch to Take Heisman Trophy."

Yet, none of it seems so real as the flood of memories that always engulfs me this time of year.

The memories about Toccoa, THS and those purple-shirted heroes of mine (my dad to coach Anderson in the equipment room: "I know we need new helmets, Eddie, but can't we just get these re-conditioned?" . . . "Sure Doc's gone, but we still got Red and John" — talk at the Dixie Lunch of Howard Ayers, Red Simmons and John Terrell as the 1941 season approached).

September would arrive at last, school open again and those magical Friday nights would come to the athletic field across from the old jail.

Coach Anderson, perspiring, always with a tie, an open collar, sleeves rolled up, marking off the field with a lime that clung like native red clay to the gray game pants.

The players, our town heroes, filing out of the basement of the old junior high building after practice, limping slightly. There'd be those purple shirts in the washwomen's backyard, starched and resplendent — a signal that Friday night would be something special.

Number one belonged to the mercurial Doc Ayers; number two, John Terrell, lanky and with nothing on his mind but football; number three, Red Simmons, jaw outthrust, a Stonewall Jackson to Terrell's R.E. Lee.

Then there were the townspeople, fiercely loyal, loving the game, but caring deeply for the players who never quite realized it was just a game.

In 1940, during a hotly-contested game, there was Doc Ayers taking a swing at an opposing player.

I tagged along as my father marched into the locker room after the game. Howard was "invited" to join us at a place called "Cinciola's for a cup of hot chocolate."

After a little small talk, old number one ducked his head and blurted, "Mr. Cheatham, I won't ever do that again." A successful coach Ayers would one day invite my father to join him on the bench as his charges took a region championship.

Too soon America was at war and Toccoans were collecting tinfoil and grease for the war effort.

Male teachers were in short supply and Dad stepped in to serve as interim football coach.

Down the road in Athens, a dream team from the University was Rose Bowl bound.

My brother, J.B., and his friend Richard Acree, could talk of little else. Richard (number 15) and J.B. (number 14) saw themselves as the "Racehorse" Davis and "Fireball" Frankie Sinkwich of the North Georgia Football Association.

They claimed the title that year, only to lose a playoff game with Hartwell. But, it seemed for a while, at the season's start, that J.B. was to be relegated to the less-than-glamorous position of center on the team.

That changed when a delegation of players came to the coach and pled that he be installed at tailback. The coach reluctantly consented to the position change, but the father was never prouder of a son who became a team leader.

It's August in Ohio. They love the football here; 87,000 partisans will jam the cavernous stadium for every Ohio State home game.

But, I have something just as precious as these Ohioans, for all the national championships and Heisman Trophy winners — a love for the game as it was played by those scrawny, but determined boys of the high school that closed in 1974.

Soon, the local high school stadium will be aglow and the cheers for the Bexley Lions will thunder down Brentwood Road.

I'll not be there, though. I'll be in my backyard hearing echoes from a gold clad THS band. I'll see a team of about 22 boys, dreaming of glory and another victory for the school that is no more.

Fall 1983

Quotable Notables

(Baseball's Greatest Quotations, by Paul Dickson)

Contrast a dictum of the late Commissioner of Baseball A. Bartlett Giamatti and quotes from many of those whose words are recorded in this bulky volume. From a literary standpoint, of course, the latter are found wanting.

It was the one-time Yale classics professor and university president who posited that "All sport aspires to the condition of paradise." He would add that it is "a dream of ourselves as better than we are."

None of the Yogi Berraisms contained herein attempts such sweep and grandeur. Worth sharing, they are. For example after laying down a comic book, the catcher says to infielder Bobby Brown, later an M.D., who was reading Gray's Anatomy: "How did yours come out?" And the same character was asked after seeing his first streaker if the person were male or female. "Don't know," answers Berra, "they were wearing a bag over their head."

Included here too is the observation of Hertford, North Carolina, native, pitcher Catfish Hunter, on slugging superstar Reggie Jackson and his "unselfishness." Said Hunter, "He'd give you the shirt off his back. Of course, he'd call a news conference to announce it."

Nevertheless, these are the stuff of which "hot stove league" discussions are made.

Some quotations from the lesser lights are poignant — or unintentionally funny — as well.

Walt Whitman, Longfellow – even Woody Allen – they're all here. But most die-hard fans will appreciate the wisdom of hellacious-good-fellows like pitcher Tug McGraw. When asked how he planned to spend his $75,000 salary, McGraw replied, "Ninety percent I'll spend on good times, women and Irish whiskey; the rest I'll probably waste."

Locker room humor, as well as the literary paeans to the national pastime, they're both here. That's quite alright with this reviewer. But, I mean, isn't the central concern whether the Atlanta Braves can repeat in the National League West?

Fall 1991

Joe Turbidy

Joe Turbidy's (L) warmth and love for youngsters who played the game will long be remembered by those whose lives he touched. A three-sport, four-year athlete at Holy Cross, Turbidy moonlighted at early NFL games, often earning as much as $10 per contest. Here he is pictured with Georgia's Wally Butts and Georgia Tech's Bobby Dodd.

"Yes," Louise Turbidy told my mother, "Joe was awfully nice to live with." Mr. Joe Turbidy's widow was right on the mark with that assessment, I'm sure.

The friendly and garrulous Irishman also was also like a second father to a sizable following of young boys in the old hometown, myself included. He seemed to us the kind of father figure represented by actor Pat O'Brien, who was cast in the Knute Rockne role in "The Spirit of Notre Dame" movie. That, of course, was the film that featured a very young Ronald Reagan as halfback George Gipp.

Remember? "Let's win one for the Gipper!"

"M-A-H-T, come see this," Joe would call to son Martin in his Irish brogue when I was over to watch sports on one of the town's first television sets. This Massachusetts native had played some good football at Holy Cross — at that time very big-time indeed. And, Lordy, could he tell a story.

There was ol' Joe, winning an argument with Illinois' immortal Red Grange, "The Galloping Ghost." On another occasion, trying to obtain a football grant-in-aid for a local lad from his friend, Coach Frank Leahy of Notre Dame.

And there was my favorite, the one about Joe playing across from a one-armed Irish guard at South Bend. "Nubbed me to death all afternoon, what a beating!"

Then the next afternoon when he suited up for pro football's Providence Steamrollers against the Canton Bulldogs. Both were then NFL teams.

The guard he faced that next day was another one-armed fellow "nubbin' me to death just like the last one." Joe needed the money. He was the son of a janitor whose mother died when Turbidy was six or seven.

Football was his ticket out of poverty. And in time he met and wooed a comely Southern belle, Louise Berry — a lady whose father just happened to be the wealthiest man in our town.

Recognizing Joe's epic gift of gab, the Holy Cross yearbook observed, "Generally speaking, Joe is generally speaking." And as often as not he was the emcee for the touchdown club jamborees and our school's team banquets.

Oh, as for the Canton Bulldog with the nub for an arm, "Say, fella, don't I know you from somewhere?" Joe asked. "Hah, yeah, I'm the one that beat you to a bloody pulp at South Bend yesterday."

"Son of a gun," Joe says, "How much they payin' you?" The fellow says, "Twenty-five every Sunday." "Dang," says Joe, who was pulling down a measly ten spot from the Steamrollers.

I would be envious when the by-then-prosperous and successful Joe took sons John and "MAHT" to the Army-Navy game or the World Series. But the man really was one to look out for the underdog; say, the undersized scrub cornerback making the best of his meager talents to break through and down the 219-pound-All-State fullback for a loss.

It was like making All-State yourself when Joe would put his beefy arm around your shoulder and say, "Good hit, pal, lots of guts!" Years later, too, I returned from making a speech to a civic gathering back home. (By then I was being billed as "a rising young corporate executive.") There his letter was, waiting for me at home. "Big-time performance, kid, big time; we're all proud of what you're making of yourself."

That felt better than if Joe had let me go with him and his boys to the Army-Navy game.

What a guy, and wouldn't you just know, there's a couple or more a lot like him in and around a good many of our towns. God bless 'em, I say. There's more than one young boy out there who hungers for their approval and their encouragement . . . in the same way as did those young guys whose lives Joe Turbidy touched four or five decades ago.

What memories those, from that somewhat sleepy little city tucked between the mountain ranges and far away from those impersonal and uncaring big cities.

Fall 1991

Turkey Day Heroes

It's just nine days away, my favorite holiday. Already the theater of my mind is playing a full schedule of re-runs around the clock. And I can hardly wait for Thanksgiving Day 1991 just ahead on November 28.

When I was growing up, each and every Thanksgiving brought a very special event: The Turkey Day Football Game. Those games are among my favorite "re-runs."

My dad, the school superintendent, had worked with community leaders to bring about consensus on the time for the traditional contest.

Nothing was to interfere with the traditional holiday dinner at mid-day, so kickoff always would come at 10 a.m. After all, the rivals were separated only by 17 miles, a modest mountain range, and the county line. And when we won — as we usually did when playing at home — our families had that much more for which to be thankful.

In those days before Pearl Harbor, life was simpler and heroes were, well, more heroic. Tradition held sway and family life was central to everything we experienced.

One of our rivals always had a Dowis boy as their star, it seemed. The opponent on Turkey Day had as its great leader the magnificent Cliff Kimsey and later his son, Bucky. Us, we had a different hero emerge each season — boys with nicknames like "Hardrock," "Whitey," "Skeets," and "Doc."

The helmets were leather, topped with flaking purple and yellow paint. The game pants sported a big purple stripe up the backside of the putty gray canvas.

By kickoff time, several hundred spectators ringed a field dotted with dying brown Bermuda grass. Just to the rear of the rowdy ranks of partisans were parked almost all the town's automobiles.

The opposing players could hear and feel each taunt, each insult the phalanx of fans hurled at them. Our boys could hear and take to heart such injunctions as, "Skeeter, son, yo' Momma and me's proud a' you; don't you let that big ol' boy back you down!"

The sheriff fed the prisoners turkey and dressing from John Sosebee's Dixie Lunch . . . and even permitted them to attend the game. Fact is, only one of those released to cheer the team on made an escape. He was the village idiot, Hambone McCluskey, who on such a furlough made straight for the depot and hopped a freight.

Hambone was headed for Greenville and freedom, where he made his home for the rest of his days. Dad said it was a bargain, seeing as how the act served to boost the average IQ of each city.

I don't remember sitting down during those games, and I couldn't imagine anyone's staying home. For us, were these games big? Oh, bigger than the Rose Bowl. Even bigger than some of the set-to's down at The State University.

Why, down there every squad member had his own red game jersey. On the other hand, we had just a dozen or so purple and gold game jersies. The eight or nine subs wore white practice sweatshirts. And if a sub was to enter the fray, the starter rapidly shucked off the jersey and his replacement ran onto the scrubby field, yanking the sleeves into place.

Then, too, each home team had its own favorite officials. One of them came up to Daddy after an especially exciting come-from-behind-win – he was coaching again that year – and said, "When your son broke that big one for the touchdown, Coach, just no way I could get that cheap Japanese whistle to blow a'tall!"

On Turkey Day the team was the town. And the town was the team. Back home after the fracas, the big logs on the fire crackled away and we inhaled the aroma of turkey and dressing coming from the kitchen.

Our family was part of the game and everything about it. Then, after Dad said the blessing, we were at one, each with the other. We were thankful in every possible way.

But too soon the war came, and our boys went off to fight. Some never came home. Me, I grew up and never saw another of those old-time contests on Turkey Day. I guess the world just changed on me.

Momma and Daddy are gone now, and the old school was shut down in 1974. But I'll be calling my brother out in Saint Louis, and we'll laugh again about The Big Game when Uncle Hotshot slipped the ref a whistle that just wouldn't blow.

It's all so much for which to be thankful. Yes, I loved those Thanksgiving Days when America was innocent and we walked among real heroes every day.

Oh, I remember those days, just like they were . . . only yesterday.

Fall 1991

Thanksgiving Day

I'm thankful for . . .

Heroes who show us how far we can go; people like Charlie Justice, Irish Eddie Hamilton, and Bobby Bell . . . John Henry Moss and the late Smoky Burgess, along with them . . . for a friend, who complimented us on a column, saying it was — a good word here — "affecting."

For Valerie Lemoignan of the athletic program at UNC-Asheville and Judy Rose, athletic director at UNC-Charlotte, who are inspiring dozens of women athletes just beginning to realize their immense potential. . . . And, yes, for the prep athlete who showed up at Lemoignan's office, saying, "I want you as my college coach — I read about you in that column in *The Star.*"

For clergy, nurses and doctors, telephone operators: All those who work through Thanksgiving Day, helping us enjoy the day off that much more. . . . And don't forget the food handlers at those few places that open up on Turkey Day.

Yep, I'm thankful for Irish Eddie Hamilton, who remembers — if few others do — those four TD passes he threw in that Turkey Day contest four decades ago. (I do, Ed.)

I'm thankful for an only child, a fiercely competitive 16-year-old daughter, who wouldn't let her dad give up when some folks thought he had gone down for the count. (Yes, the self-same one who told me "and I never WILL give up on you, Daddy.")

To the *Atlanta Journal-Constitution's* Furman Bisher, who wrote the father ready to throw in the towel. "Your letter just doesn't sound like the Mike Cheatham I knew!" . . . For Roger over at Gardner-Webb — who'd be embarrassed if I gave his full name — the finest Christian I ever knew, day in and day out.

The turkey and dressing on Thanksgiving Day, even if it is at the Holiday Inn. . . .

And, of course, for the candy corn they served at my church in Dallas in place of crackers for the communion service that wonderful day in 1973. . . . For Possum and Dank, whose, er, aroma was so bad that no one invited them in for turkey back in my old hometown. And for the tenderhearted deputy who always would put out plates for them in back of the old jail on the big day.

I'm thankful too for Rand, my old bachelor friend who does things like serve Thanksgiving dinner at the mission on the day; he just may be the noblest of my heroes. . . . And for the late Commissioner of Baseball A. Bartlett Giamatti, who reminded us that all sport "aspires to the condition of paradise" and that it is "a dream of ourselves as better than we are. . . ." For his words and his inspiration, may we all be thankful.

For Yankee transplants Joe Harand of Shelby and Jeff Renk of Asheville, who know our strange, xenophobic ways — and still care for us. . . . For Alan, Michael, and David on *The Star* sports desk, who taught an old broadcast type how to write for print. . . . And for the docs who made our editor Tom, well — and a lot of others of us really happy, as well.

For newly-found friends who let me use (read that, misuse) their word processor to turn out this column. . . . And for John and Carol Schramm, never judgmental of others, who give hope and sustenance to so many on every day in the year.

And, Lord, thank you for a day long ago when Coach Ed White made it possible to wear my first and only letter in front of all the family on Turkey Day 1947 — For that, and most of all, for a mighty good momma and daddy.

Now, will you do just one little thing for this foolish old writer, dear reader? Will you just share a little of that white meat today with someone who didn't quite make one of life's many cuts this year?

A few kind words and a little of Momma's giblet gravy might be appreciated, too.

Fall 1991